Deep-Sea Minerals Developments
in the 20th Century

David S. Cronan

Deep-Sea Minerals Developments in the 20th Century

 Springer

David S. Cronan
Emeritus Professor
Royal School of Mines
London, UK

ISBN 978-3-031-52341-0   ISBN 978-3-031-52342-7   (eBook)
https://doi.org/10.1007/978-3-031-52342-7

This Springer imprint is published by the registered company Springer Nature Switzerland AG
The registered company address is: Gewerbestrasse 11, 6330 Cham, Switzerland

Paper in this product is recyclable.

Preface

From originally being thought of as a good idea to help overcome expected mineral shortages, deep-sea mining has become a subject of considerable controversy. This book describes deep-sea mineral activities during much of the period that this change of heart took place and aims to provide some historical background to the current debate. In writing it, I have drawn extensively on personal experience, conversations with individuals working in the field, the open literature (including the Ph.D. theses of some of my past students), and conferences and meetings attended. Because much of the twentieth-century material on deep-sea mineral-related activities never appeared in library-held peer-reviewed journals, but in trade journals, government and industry reports, and at conferences, I have had to draw extensively on these. Chief amongst the conferences was the Underwater Mining Institute held annually from 1970 until it became the Underwater Mining Conference in the second decade of the twenty-first century. It was founded and chaired by J Robert (Robby) Moore from its inception until Robby's death in 1995. I was a regular attendee from the 1970s onwards. Robby prided himself on being at the forefront of knowledge dissemination on developments in deep-sea mineral-related activities and was able to get speakers at the cutting edge of research. "You heard it first at the UMI" was an often-used catch phrase of his. Much of the material covered in this book was first aired at a UMI, and extensive use has been made of this source. Of course, other conferences were also of importance in this regard, such as the IDOE 1972 Arden House Conference, several conferences at the University of Hawaii and the East-West Centre from 1973 onwards, and the frequent CCOP/SOPAC conferences in the SW Pacific from 1975 onwards. I was fortunate to have attended many of these, and their proceedings (published and unpublished) feature prominently in this book. Almost all the material is pre-Internet.

London, UK   David S. Cronan

v

Acknowledgements

I would like to express my gratitude to all who have helped me in completing this book, either by providing information via personal communications (acknowledged in the text) or by reading and commenting on chapters or parts of chapters. I would also like to thank those organizations with whom I have had the privilege of working, including the University of Hawaii, the New Zealand Oceanographic Institute, CCOP/SOPAC, and the International Seabed Authority, experiences that strongly influenced my thinking on much of the material in the text.

General Introduction

The concept of mining minerals underwater dates back to at least the eighteenth century. Lake ferromanganese oxide concretions had been known and mined in the northern Baltic region in the middle ages. Apparently, these grew so fast that the lakes were "cropped" every so many years thereby providing a renewable source of the minerals. More recently, similar activities were carried out in eastern Canada.

Manganese nodules were the first deep-sea minerals to be recognized and were first discovered in the Atlantic Ocean during the Challenger Expedition (1873–1876) around 300 km SW of the Canary Islands on February 18, 1873 [1]. During the Challenger Expedition, large numbers of nodules of varying natures were recovered from the three major oceans. They mainly consisted of concentric bands of ferromanganese oxides around nuclei of various types. Some of the nodules had been fractured and new ferromanganese oxides accreted on the broken surfaces. The morphology of the nodules often reflected the shape of the nucleus, and sometimes multiple nuclei were present.

It was in the Pacific that HMS Challenger found the largest nodule deposits, on red to chocolate-coloured clays. Chemical analysis of them showed that their major constituents were iron (Fe) and manganese (Mn). Minor elements were found to include copper (Cu), nickel (Ni), and cobalt (Co), the main elements of twentieth-century economic interest in the deposits. Many other elements now known to occur in nodules were beneath detection by the analytical methods in use in the late nineteenth century. Glasby [2] has provided more detail of the early history of manganese nodule discovery and pointed out that it was the Challenger Expedition chemist, JY Buchanan, who, in a letter to his father, first drew attention to the commercial possibilities of the nodules.

According to Glasby [2], following the Challenger Expedition, work on manganese nodules was sporadic. Collections were made from the Pacific during the Albatross Expeditions of 1899–1900 and 1904–1905 and the Carnegie Expedition of 1928–1929. Nodules were also recovered from the Indian Ocean during the Valdivia Expedition of 1898 and the John Murray Expedition in 1933–1934. They were also collected on the Swedish Deep-Sea Expedition in the late 1940s.

Ferromanganese oxide crusts were not always distinguished from nodules in the early days. Using dredging for recovery, as was almost always the case, crusts and nodules could get broken and mixed up in the same dredge. As will be outlined later, they were finally recognized as a separate class of deposits in the 1980s.

Phosphorites, being mainly continental margin deposits, are not universally recognized as deep-sea deposits. They were found in the South Atlantic off southern Africa in the early years of the twentieth century and in many other locations since [3].

Deep-sea hydrothermal deposits were first recognized in the Red Sea in the 1960s [4], although anomalous bottom water temperatures had been recorded there in 1948 during the Swedish Deep-Sea Expedition.

Early work on deep-sea minerals was mainly scientific. It was not until the 1950s that significant thought was given to their commercial potential. The possible effects on the marine environment of mining them were rarely considered. This is in marked contrast to the present day when environmental considerations are at the forefront of deep-sea mineral-related activities. Much of the present controversy surrounding deep-sea mining has its roots in the late twentieth century.

References

1. Murray J and Renard AF (1891) Deep-sea deposits. Report of the scientific results of HMS Challenger, 1873–1876. HMSO, London
2. Glasby GP (1977) Marine manganese deposits. Elsevier, Amsterdam
3. Cronan DS (1980) Underwater minerals. Academic Press, London
4. Miller AR, Densmore CD, Degens E et al (1966) Hot brines and recent iron deposits in deeps of the Red Sea. Geochim Cosmochim Acta 30:341–359

Contents

Part I   Initial Enthusiasm,1957-Early 1980s

The Post-Second World War Deep-Sea Minerals Scene   3
Introduction.   3
Positive Assessments of Deep-Sea Mining and Why It Might Be Needed. . .   4
Negative Assessments of Deep-Sea Mining and Why It Might Not Be
    Needed   8
External Uncertainties.   9
Environmental Issues   10
Law of the Sea and Deep-Sea Mining   11
References.   14

Activities on Manganese Nodules During the Post-war Boom   17
Introduction.   17
Pacific Ocean   17
    North Pacific Ocean   20
    South Pacific Ocean   23
    Transects   25
Atlantic Ocean   30
Indian Ocean   31
Resource Potential   32
Economic Evaluations.   34
Postscript: Project Azorian, the Glomar Explorer Affair   39
References.   39

Cobalt-Rich Crusts: Recognition and Preliminary Evaluations   43
Introduction.   43
Regional Studies   43
Crust Characteristics   46
Economic Potential   46
References.   47

Hydrothermal Deposits: Discovery and Preliminary Economic
  Evaluation 49
  Introduction. 49
  Red Sea Deposits 50
  Polymetallic Sulphide (PMS) and Related Deposits 52
  Metalliferous Sediments 54
  Hydrothermal Ferromanganese Oxide Crusts. 55
  References. 56

Phosphorites 59
  Introduction. 59
  Occurrence and Associations 59
  Resource Potential 61
  References. 62

Exploration and Mining Development 63
  Exploration 63
    Visual Methods 63
    Sampling 64
    Remote Methods 65
    Geochemical Methods 67
  Mining Development 68
    Manganese Nodules 68
    Hydrothermal Deposits. 71
    Co Rich Crusts 72
  References. 72

Part II   Transition, Circumspection and Diversification,
         Early 1980s–2000

Hydrothermal PMS and Related Deposits 77
  Spreading Centres. 77
  Convergent Plate Boundaries 79
  End of Century PMS Activities. 81
  References. 83

Expanding Cobalt-Rich Crust Activities in the Pacific Ocean 85
  Introduction. 85
  North Pacific 85
  South Pacific 88
  Resource Potential 89
  References. 92

Late Twentieth-Century Manganese Nodule Activities 95
  International Seabed Area. . . . . : . . 95
    North Pacific 95
    South Pacific 99

Indian Ocean.   100
Exclusive Economic Zones.   102
  Blake Plateau   102
  SW Pacific.   102
References.   104

Technological Developments 1980–2000.   107
Introduction.   107
Seabed Mapping, Surveying and Photography.   108
Seabed Sampling.   109
Hydrothermal Plume Detection   110
References.   110

Environment   113
Introduction.   113
Manganese Nodules   114
Post-DOMES Benthic Impact Investigations   115
  Benthic Impact Experiment   116
  Japan Deep-Sea Impact Experiment.   116
  Interoceanmetal (IOM) Benthic Impact Experiment   117
  Disturbance and Recolonization Experiment (DISCOL)   117
  Indian Deep-Sea Environmental Experiment (INDEX)   118
  Synthesis.   119
Polymetallic Sulphides   119
References.   120

Economic and Legal Developments, 1983–2000.   123
Introduction.   123
Hydrothermal Minerals.   124
Manganese Nodules   125
Cobalt-Rich Crusts   130
Law of the Sea   132
References.   134

Epilogue   137

References   141

Index.   143

List of Figures

Part I   Initial Enthusiasm,1957-Early 1980s

Fig. 1   Mid-20th century metal prices.   2

The Post-Second World War Deep-Sea Minerals Scene

Fig. 1   Location of the Clarion-Clipperton Zone (CCZ) and other
          areas of potentially economic manganese nodules in the central
          Pacific Ocean   5

Activities on Manganese Nodules During the Post-war Boom

Fig. 1   Manganese nodule compositional zones in the Pacific
          Ocean, as perceived in 1972   19
Fig. 2   The Hawaii-Tahiti Transect, 1978–1980 and detailed study
          areas on it occupied by RV Suroit and RV Sonne   26
Fig. 3   The Wake-Tahiti Transect (1980) and detailed study
          areas on it as compiled by the Geological Survey of Japan
          in the late 1980s.   28
Fig. 4   The Aitutaki-Jarvis Transect (1987) and station groups
          occupied on it by the RV Thomas Washington   29
Fig. 5   Salient chemical characteristics of Indian Ocean manganese
          nodules as perceived in 1972.   31

Cobalt-Rich Crusts: Recognition and Preliminary Evaluations

Fig. 1   Areas of Co-rich crust investigations in the Hawaiian
          Archipelago in the early1980s.   44
Fig. 2   Location of the MIDPAC 81 cobalt-rich crust expedition
          showing sample locations II to XIII. (Courtesy of P Halbach)   45
Fig. 3   Sample of cobalt-rich ferromanganese oxide crust.
          (Courtesy of JR Hein) [Crust 8/10]   46

Hydrothermal Deposits: Discovery and Preliminary Economic Evaluation

Fig. 1   Location of brine pools and metalliferous sediments
          in the Red Sea in 1975   51
Fig. 2   Known and predicted locations of hydrothermal deposits
          in the SW Pacific in 1981   55

Phosphorites

Fig. 1   Worldwide continental margin phosphorite occurrences   60
Fig. 2   Area of phosphorite occurrence on the Chatham Rise   61

Exploration and Mining Development

Fig. 1   Free fall camera trigger weight and Pacific seafloor
          manganese nodules, 1986   64
Fig. 2   Free fall grab for manganese nodule sampling, 1987   66
Fig. 3   Box corer with manganese nodules at the sediment
          surface, 1987   66

Hydrothermal PMS and Related Deposits

Fig. 1   Known worldwide submarine hydrothermal occurrences
          in the late 1980s   77

Expanding Cobalt-Rich Crust Activities in the Pacific Ocean

Fig. 1   Settings of Co-rich crusts on seamounts, guyots
          and island slopes   86
Fig. 2   Locations of Co-rich crust studies in US Pacific territories
          up to circa 1990, with average Co values of the crusts
          obtained in each territory   86
Fig. 3   Nations participating in the Japan-SOPAC program, 1985–2000 . . .   88
Fig. 4   Schematic of operational facilities on Hakurei-Maru No 2   89
Fig. 5   Cobalt-rich crust thicknesses in the SOPAC region   90

Late Twentieth-Century Manganese Nodule Activities

Fig. 1   Known distribution of manganese nodules in the oceans
          circa the mid-1980s   96
Fig. 2   National consortia exploration claims in the CCZ. (Source ISA) . .   97
Fig. 3   World EEZs and International Seabed Area   99
Fig. 4   Manganese nodule sample sites in the Indian Ocean
          up to the early 1980s   101
Fig. 5   Manganese nodule abundances in SOPAC member countries   103

Economic and Legal Developments, 1983–2000

Fig. 1   Metal prices at the end of the twentieth century and into 2021   124

Epilogue

Fig. 1   REE prices in the first decade of the twenty-first century   138

About the Author

David S. Cronan   Has been working on and writing about deep-sea minerals since the mid-1960s. He completed a dissertation on them at Oxford University in 1964 and a Ph.D. on them at Imperial College, University of London, in 1967. He was awarded a D.Sc. by Durham University in 1986 or his marine minerals publications, and an Honorary D.Sc. by the University of the Aegean in 2002. From 1973, he was successively Lecturer, Reader, and Professor of Marine Geochemistry at the Royal School of Mines, Imperial College, where he is now an Emeritus Professor, working principally on manganese nodules and hydrothermal deposits. During this period he supervised over 20 Ph.D. projects on deep-sea minerals, published more than 150 papers and 4 books, and attended more than 50 conferences on them. Many of the last of these have provided material for the present book.

Previous books on deep-sea minerals by the author include

 (i) Cronan D S . Underwater Minerals, Academic Press, 1980.
 (ii) Cronan D S (ed). Sedimentation and Mineral Deposits in the Southwestern Pacific Ocean, Academic Press, 1986.
(iii) Cronan D S. Marine Minerals in Exclusive Economic Zones, Chapman and Hall, 1992.
(iv) Cronan D S (ed). Handbook of Marine Mineral Deposits, CRC Press, 2000.

Part I
Initial Enthusiasm,1957-Early 1980s

Introduction

The period of initial enthusiasm for deep-sea mineral exploration and recovery more or less coincided with the post-Second World War boom and was largely a consequence of it. That boom is generally reckoned to have lasted from about 1950 to the mid-1970s. Most of the initial advances in deep-sea mineral-related activities took place during that time.

In the post-Second World War period there was a need to find new sources of metals. This was partly in order to replace the stock that had been used up during the war, and partly, in developed countries, to provide consumer goods as their populations and economies expanded. Another factor of importance in this regard was the decolonization of large parts of Africa and Asia. These had hitherto supplied minerals on preferential terms, but now had the power to increase prices or withhold the minerals altogether. A perceived political instability in some of the newly independent countries, with the consequent possibility of mineral supply disruptions, also encouraged marine minerals activities by the developed nations in the 60s and 70s. Increasing prices for Ni and Co throughout the 50s,60, and 70s (Fig. 1), two of the three main metals of economic interest in manganese nodules, also stimulated interest in deep-sea mining. These original metals of economic interest, together with Cu, are still of economic interest today but have been augmented by a number of other elements of use in the newly developing technology industries.

Engineering and instrumental developments also supported marine minerals development. Many of these developments had taken place during the war. Examples include improvements in navigation and bottom surveying utilizing radar and echo sounding. In the case of the latter, the development of the single-beam Precision Depth Recorder (PDR), ubiquitous in seafloor exploration during the 60s and 70s, was crucial to the development of deep-sea mineral exploration. Ultimately, multibeam instruments such as Seabeam, which could provide a real-time bathymetric map of the seafloor on either side of a survey ship whilst underway, aided deep-sea mineral exploration considerably. New developments in sea bottom sampling techniques such as free-fall grabs and corers were also important during this period.

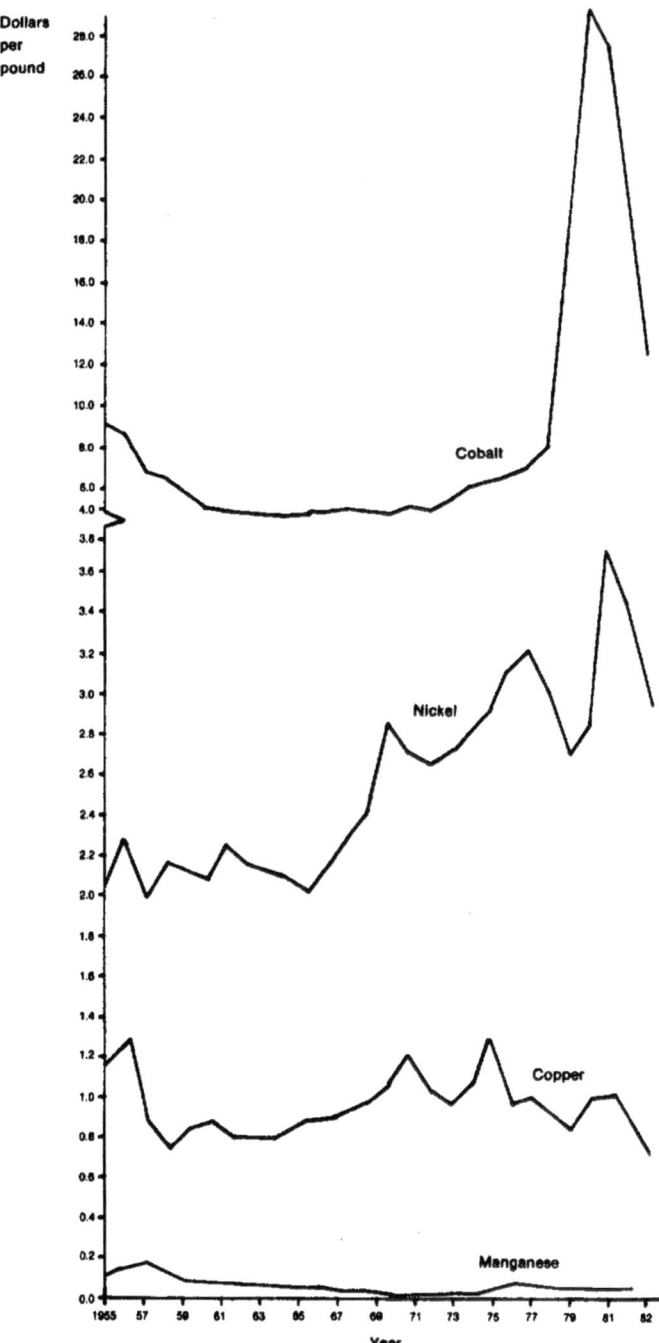

Fig. 1 Mid-20th century metal prices

The Post-Second World War Deep-Sea Minerals Scene

Introduction

Economic interest in deep-sea minerals commenced in the late 1950s. According to Mero [1], in an attempt to capitalize on the perceived economic need for new sources of metals from the sea, in late 1957 several University of California Professors met at the Scripps Institution of Oceanography to plan a cooperative program between two of the University departments (Marine Technology at Berkeley and the Institute of Marine Resources at Scripps) on the recovery of minerals from the sea floor. As the subject of the first project, they chose a study of manganese nodules on the Pacific deep-sea floor. A report came out of this project [2] which suggested that it might be possible to mine manganese nodules as a source of various metals. The program was then broadened and developed into a study of other possible mineral resources on the sea floor [1].

An early attempt to put marine minerals into the perspective of minerals at large was by Tooms [3]. He emphasized the important point that marine minerals should not be viewed in isolation but be considered in terms of the economics of competing land-based sources. He also made the point that in addition to the more expensive nature of marine-based as opposed to land-based mining operations, much basic information was supplied to land operators at no cost by Government agencies like Geological Surveys, something that was not available to potential marine miners. This increased the cost differential between the two types of operations. For this reason, he made a plea for Government assistance in obtaining basic marine geological and oceanographic information to assist in marine mineral surveys, a plea that largely fell on deaf ears. Tooms noted that there were few if any countries in which the full mineral potential had been developed as a sole result of individual mining company activity, and that it was unlikely that such would occur on the sea floor.

In a comprehensive review, Dunham [4] noted that during the 1960s the United States was the most substantial participant in marine minerals activities. He reported

D. S. Cronan, *Deep-Sea Minerals Developments in the 20th Century*, https://doi.org/10.1007/978-3-031-52342-7_1

that there was by 1965, a widespread and intense controversy in the United States "concerning the national effort to explore, understand and develop the oceans" as a result of which the President set up a panel on oceanography [5]. One of its conclusions was that a division of effort between government, industry, and universities would be advisable in the oceans, unlike in space exploration which was then entirely in government hands. This set the pattern for US marine minerals activities for decades. Nineteen sixty-six was a significant year in the United States for marine minerals activities as it saw the passing of the Marine Resources and Engineering Development Act which developed a new national policy [6] " to intensify the study of the sea and to convert to practical reality its inherent promise for man's benefit". Later while the US Government's contribution to the search for deep-sea minerals was significant, mainly through the USGS and the Bureau of Mines, it was nevertheless only a small fraction of the total expenditure on this work. Industry provided most of the rest [7].

Positive Assessments of Deep-Sea Mining and Why It Might Be Needed

Mero [1] founded his assessment of the potential worth of deep-sea minerals principally on manganese nodules and the situation in regard to mineral supply and demand in the U S A in the 1950s and 1960s. Mero pointed out that Americans had increasingly looked to foreign nations for supplies of raw materials. He believed that because of increasing populations and rising standards of living in those nations, with a concomitant rise in mineral consumption, those sources were beginning to become less reliable. From this, he concluded that there remained few areas on the planet to seek new sources of minerals other than the sea floor. He believed that eventually, political and population pressures would force the more highly industrialized nations into exploiting the sea floor for several metals including Ni, Cu, Co, and Mn, and further in the future possibly V, Pb, Zn, and REE. Nevertheless, Mero believed that it was unlikely that mankind would turn to deep-sea minerals because those on the continents had been used up, but because it would be cheaper to mine them from under the sea than from on land. He assumed that if only 10% of the nodule deposits could be mined economically, there were sufficient supplies of many metals in them to last for thousands of years at the then rates of consumption. Also, he pointed out that, on a global scale, some elements were accumulating faster in the nodules than they could be utilized at the then rates of consumption, three times as fast in the case of Mn, four times as fast in the case of Co, and as fast in the case of Ni [8]. However, this did not take into account that the metal richest, most likely to be mined, nodules are among the slowest growing and would not grow faster than they could be recovered. Mero did point out that there would be marketing problems in disposing of some of the metals (but not of Cu) from the nodules if they were mined at a rate to produce Ni (the main metal of interest) economically.

However, he established that nodules varied markedly in composition from place to place on the ocean floor and thus mine sites could possibly be chosen to recover nodules that allowed most of their metals to be disposed of without seriously disrupting the metals markets. It was the highly optimistic projections of Mero that were most contentious. However, his main achievement remains intact, namely that he successfully outlined the different compositional regions of manganese nodules in the Pacific based on very limited data, regions that have remained more or less intact with the gathering of more data over the ensuing years.

Feeling that his work, as presented in his 1965 book, had been misunderstood, or even misrepresented, Mero sought to clarify his position on deep-sea mining [9, 10]. He reiterated his basic position that if everybody on earth was to approach a standard of living similar to that in the developed nations, raw materials other than those on land would have to be found. He pointed out that ocean mining had seen some spectacular publicity in the decade since the publication of his book, but largely for the wrong reasons. For example, he considered that errors in calculation had led some politicians to the belief that there were hundreds of billions or even trillions of dollars worth of nodules lying out in the oceans, the mining of which would provide funds to solve many of the World's problems.

According to Glasby [11], Mero's work was largely responsible for the 1970s being a decade of intensive commercial exploration and scientific activity on nodules, the former mainly in the Clarion-Clipperton Zone (CCZ) of the tropical North Pacific (Fig. 1). In addition to the commercial activity already mentioned and to be described more fully in the next chapter, the United States declared the 1970s to be an " International Decade of Ocean Exploration"(IDOE) which became the umbrella

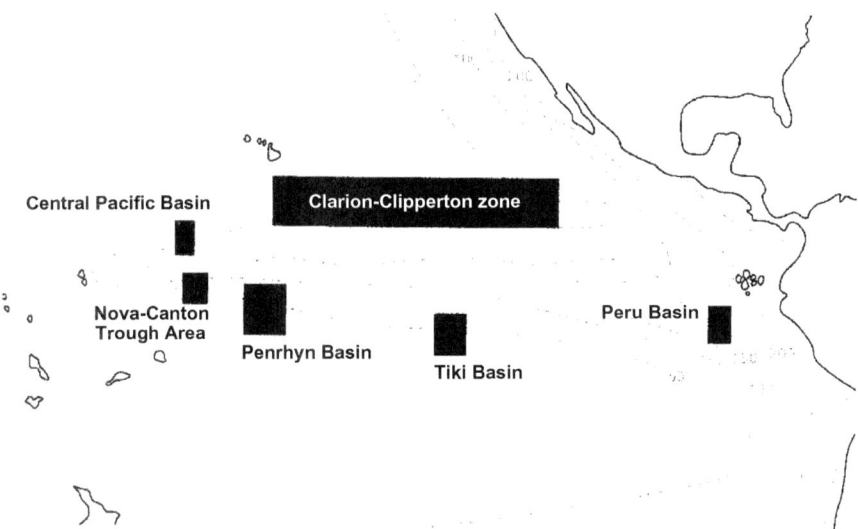

Fig. 1 Location of the Clarion-Clipperton Zone (CCZ) and other areas of potentially economic manganese nodules in the central Pacific Ocean

under which much of the non-industrial work on deep-sea minerals was carried out. A major part of this was the "Inter-University Manganese Nodule Project (IUMNP). According to Gerard [12], this was initiated in 1972 at the "Arden House Conference", which was a conference funded by the NSF under the IDOE Project at Arden House of Columbia University, and was the first of the great manganese nodule conferences which, together with successive Underwater Mining Institutes (UMIs) from 1970 onwards, did so much to spread understanding of manganese nodules and whose proceedings remain to this day important works in manganese nodule science and technology. The first phase of the IUMNP was completed in 1973 and included baseline studies covering the then state of knowledge of manganese nodules on the ocean floor [12]. One of the major tasks during this early part of the IUMNP was to plot all available data on the worldwide distribution and metal content of the nodules. This confirmed that the richest deposits lay in the eastern tropical North Pacific between 5–20 N and 115–150 W (the CCZ). The second phase of the Project began in 1974 and included three cruises in the CCZ to help understand the factors that determine the nature and distribution of the nodules. Studies initiated during 1975–1976 included investigations on the role of the particulate flux of biological material to the sea floor and its effect on nodule growth and elemental composition, and long-term observations on nodules and the sea floor environment by use of time-lapse photography, light scattering measurements and measurements of the variability in bottom currents with time. Developing from the IUMNP (which was primarily a US program), and running in parallel with it during its later stages, were the "International Cooperative Investigations on Manganese Nodule Environments" (ICIME) which involved both US and non-US institutions [13]. Among other things, this concentrated on a N-S traverse in the tropical Pacific at 134 W which investigated many aspects of nodule formation and variability in relation to their depositional environment (see Chapter "Activities on Manganese Nodules During the Post-war boom").

Although the initial interest in exploiting deep-sea minerals came from the United States, starting in the 1960s other countries found it in their interest to study deep-sea floor manganese nodules from an economic perspective [4]. From West Germany, there were cruises of R V Valdivia and later cruises of R V Sonne in the Pacific and Indian Oceans. France undertook several cruises in the Pacific and was the first to carry out submersible dives on nodule fields. The Soviet Union also undertook a large number of cruises worldwide starting in the 1960s, and in the mid -1970s the New Zealand Oceanographic Institute (NZOI) undertook two nodule exploration cruises in the SW Pacific. The UK Govt partly funded the involvement of Rio Tinto and Consolidated Goldfields in the Kennecott Manganese Nodule Consortium (see Chapter "Activities on Manganese Nodules During the Post-war boom"), but Glasby [14] believed that overall Britain had a very limited commercial interest in nodules and that he was "never able to establish whether this was due to foresight or apathy".[1] However, there was a view in Britain in the 1960s that it could possibly claim vast areas in the southern and western Pacific for mineral exploitation, if it wished, by virtue of its colonies there [15]. Of course,that never happened because all those colonies except one became independent. However, they themselves later became interested in deep-sea mineral exploration and exploitation, as will be described in subsequent chapters. According to Glasby [14], an indication of the scale of interest in deep-sea mining worldwide in the 1970s can be judged from the fact that by 1982 almost 4000 publications had been produced on manganese nodules, over 70% of which appeared in the period 1972–1982.

Subsequent considerations on the need for deep-sea mining were presented by Pendley [16]. Pendley rejected recycling, substitution, and diversification of land supplies as possible sources of sufficient minerals to make deep-sea mining unnecessary. He further stated that because the US was dependent on foreign sources for a significant portion of its nodule metals Ni, Co, and Mn, (only Cu was available domestically in sufficient quantities to support US needs) the mining of deep seabed mineral resources (nodules) could contribute substantially to the US domestic supply of these minerals.

One of the most publicized factors leading to Pendley's appraisal of US deep seabed mining needs was security of minerals supply, particularly of Co. Cobalt was a metal with an important use in heavy industry, space and defense applications (more than 900 pounds of Co were needed to produce an engine for the F-16 fighter jet), and the US was vulnerable to disruption in its supply and hence to fluctuations in its price. In 1977 and 1978 when the world's principal cobalt producer, Zaire, was invaded by insurgents from Angola and Zambia, the price of Co quadrupled (Figure 1 of "Initial Enthusiasm, 1957-early 1980s"). It came back down again after the invasion was repelled, but not immediately, causing concern in the metals markets about the effects of such actions. An additional concern in regard to possible supply disruptions commented on by Pendley was that the Soviet Union could engage in a resource war by disrupting world markets and limiting the supply of metals (it was still the Cold War at that time). Pendley's conclusion was that the critical need of the US for "strategic" raw materials, even allowing for stockpiling, substitution, and recycling, was such that deep-sea minerals could not be ruled out as a resource. According to Pendley, it was this view of America's strategic mineral interests that influenced the decision by the US to undertake a fundamental re-examination of the then-draft Convention on the Law of the Sea and the interests of the US in it (see below).

---

[1] In the writer's opinion, most likely the UK's seeming limited interest in deep-sea mining was deliberate and due to a combination of, firstly, Britain having negotiated favourable mineral supply contracts with its ex-colonial territories in the 1950s, 1960s, and 1970s, mainly in Africa, and, secondly, a UK Govt resource specialist, R D Medford, having carried out an analysis in the 1960s that indicated that deep-sea mining was unlikely to be economically viable. At a meeting on marine resources as part of the London School of Economics Sea-Use Course in the 1980s, the writer met David Anderson, a UK Foreign Office legal advisor during the Law of the Sea Conference and a member of the British negotiating team at the Conference. He told the writer that when the negotiations started, the Foreign Office was asked by the Government to make a list of 10 marine interests, in order of importance, that Britain might have in any future Convention. Deep-sea mining was number 10.

Negative Assessments of Deep-Sea Mining and Why It Might Not Be Needed

It was not long after John Mero's pronouncements on deep-sea mining that others put forward alternative views. Among the first of these was R. D. Medford of the UK Govt's Programme Analysis Unit (PAU). In the late 1960s, the UK Govt's Ministry of Technology commissioned the PAU to undertake a study of UK mari= time activities which resulted, among other things, in a report entitled Potential National Benefits from R&D in the Field of Marine Mining. Consequent on this were two articles by Medford [17, 18] in Mining Magazine. Medford outlined a mathematical model to facilitate discussion of various cost parameters of marine mining relative to terrestrial mining. Medford's analysis concerning deep-sea mining was somewhat pessimistic. He commented that metallurgical investigations carried out at Harwell and the University of California indicated that no new developments were imminent that would lead to more efficient processing by which Ni and Cu could be extracted more economically from nodules. Further, no new deep-sea recovery process was known that would allow nodules to be mined in *reasonable (his italics)* quantities. He pointed to Mero's claim that the economic feasibility of the recovery of manganese nodules depended on a very high capital investment and a throughput of 5000 tonnes per day, which he considered unrealistic. Also, he drew attention to the work of F. T. Christy of Resources for the Future Inc. who suggested that one successful marine mining operation would flood the market with metals, causing a drop of 10–40% in the price of Mn, Co, and Ni. He commented that investors appeared reluctant to finance deep-sea mining operations because of the large capitalization required to provide the large throughput demanded from then-perceived economic viability. Thus, in essence, Medford argued against deep-sea mining first because the scale of the operation needed for economic viability was too large to be feasible, and second that it would depress metal prices and thus decrease the profitability of any successful operation.

Drechsler [19] attempted to provide an economic context for any new manganese nodule mining industry. He pointed out that there was no foreseeable shortage of any of the metals of economic interest in the nodules (Mn, Ni, Co, Cu) at that time at the then-current prices, and that this was expected to remain the same up to at least the year 2000 assuming (i) that the rate of growth in consumption did not change, (ii) the relative price of competing products did not change, (iii) there was no major increase or decrease in world income and (iv) that consumers would continue to want metal goods. From this, he concluded that long-run metal prices would not rise much during the following 30 years and might even drop somewhat (which they did). He pointed out that as mining decisions were made on a long-term basis, his prognostication might sound pessimistic for deep-sea mining (which it was). However, he tried to distance himself from this view by pointing out that scientific and technological issues needed to be taken into account.

The questionable need for deep sea mining and some of its possible disadvantages was also considered by Clark [20]. Examining the four major metals of interest in the nodules at that time Mn, Ni, Cu, and Co, he concluded that as far as the USA was concerned, possible withholding of supplies of these metals by "actions of the sort that happened in Zaire and Zambia in the late seventies" would be insignificant for most of them. Only Co (and to a lesser extent Mn) was perceived to be vulnerable to such disruption. However, he admitted that supplies of Co and Mn from deep-sea nodules could reduce the expected cost of any such disruption and might eliminate the need for extensive stockpiling of them as existed in the US at that time. Nevertheless, Clark further argued that the adverse political conse-quences of the US mining deep-sea minerals outside the LOS Convention (the US had just declined to sign it at that time) could be substantial, as he noted that the great fear of the developing world was that the industrialized countries will gain control of the supply of metals gained from deep-sea nodules and that would reduce the mineral revenues of some of the developing states. Clark concluded that the national security implications of the US not mining deep-sea minerals were negli-gible and that the economic benefits of the US unilaterally mining them were insuf-ficient to make it an attractive proposition. Nevertheless, if deep-sea mining were to take place under the provisions of the Law of the Sea Convention, he believed that the benefits accruing over a 20-year period would not be negligible.

A more extreme negative position on deep-sea mining was taken by Glasby [14]. He pointed out that a key factor in initiating research on the possibility of deep-sea mining was the prediction of global mineral shortages based on forecasts of the Club of Rome [21]. Glasby pointed out that these forecasts turned out to be in error because they assumed fixed reserves of minerals on land and failed to consider that their depletion would result in an increase in the prices of the minerals concerned which would lead to recycling and exploration for new deposits of them on land. He concluded that the concept of manganese nodule mining was based on a false prem-ise and was therefore deeply flawed from the outset. He pointed out that in spite of this, from the early 60 s up to 1984 more than US $ 650 million was spent on explor-ing the deep-sea for nodules and developing technologies for mining them, with peak spending occurring in the period 1976–1980. He concurred with Broadus [22] who, based on economic considerations, believed that deep-sea mining was a mis-taken investment in "a wasteful losing venture". Nevertheless, Glasby [14] pointed out that for most of the companies involved it was a relatively small proportion of their total annual budgets, 5–7% on exploration and R&D during the peak spending period. No companies went bankrupt because of misplaced spending on deep-sea minerals development.

External Uncertainties

In 1973 an oil price crisis resulting in a sharp increase in energy prices had a damp-ening effect on progress towards deep-sea mining, and oil prices remained unpre-dictably high for much of the remainder of the decade. For example, the proposed metallurgical processing of the nodules to extract metals of interest was going to be very energy intensive. Some sources put it at more than 50% of the total cost of manganese nodule exploitation [11, 23] and thus it would benefit from cheap energy. The sharp increase in the oil price in 1973 and its effect on energy costs seriously impacted on the perceived cost and profitability of ocean mining and prompted a reappraisal of the viability of the operation. Coupled with increased energy costs a second damper on manganese nodule mining development in the mid-1970s was the uncertainty surrounding the legal regime under which the nodules might be mined. A hoped-for early resolution did not happen. According to Wright [24], by the mid-1970s some mining companies had completed preliminary investigations into locating and evaluating nodule deposits, had completed small-scale testing of mining and recovery systems, and were preparing to enter a more costly phase of pre-production development. However, industry spokespersons indicated before the US Congress considerable reluctance to commit to the level of funding needed to conduct this pre-production effort in the absence of the greater investment stability that would result from a settled legal regime. Further, according to Earney [7], some managers in the developing deep-sea mining industry even felt that the only truly viable mining regime would come from outside the LOS Convention and considered that free-market mining operations would be impossible under it. Thus by the mid-1970s, the stage was set for a period of relative inactivity in the development of deep-sea mining until at least the legal issues could be resolved and either energy prices came down or nodule processing costs could be reduced.

Environmental Issues

Mero [1] never mentioned in his 1965 book any environmental problems likely to be associated with deep-sea mining. However, an oft quoted point that he did make was that deep sea-mining could lead to the closure of polluting land mines. According to Mero [10], the US Govt commenced interest in the possible environmental effects of mining deep-sea minerals in 1969 and tasked the National Oceanic and Atmospheric Administration (NOAA) of the US Dept of Commerce to sponsor several studies on the environmental aspects of nodule mining. In the 1970s the possibility that deep-sea mining might cause environmental damage started to be taken seriously and this led to the DOMES Project.

In the late 1970s (1975–1981), Project DOMES (Deep Ocean Mining Environmental Study) was concerned with obtaining baseline data from the CCZ. Much of the initial work was done in the photic zone where discharges from nodule-lifting devices could occur. According to Earney [7] any system for lifting nodules through the water column would release large amounts of sediment into it. From extrapolation of data obtained during a prototype mining test in 1978 it was estimated that discharged wastes in a full-scale mining operation could create a sediment plume up to 100 km in length and 10–20 km wide, and could cause substantial reductions in light levels in the affected area leading to decreased photosynthesis [25].

According to Lipton et al. [26] and Parianos [27], Part I of the program called DOMES I was undertaken by NOAA to provide environmental baseline information on three representative mining sites (A, B, C) in the Pacific manganese nodule mining province in and near the CCZ. Each covered an area of approximately 200 km by 200 km and was chosen in consultation with industry and the scientific community. Twelve cruises of NOAA's research vessel R/V Oceanographer were carried out from August 1975 through November 1976, totaling approximately 240 ship days at the three sites. Scientific disciplines represented were physical oceanography (studies of solar radiation and ocean currents), biological oceanography (studies of phytoplankton and benthic fauna), chemical oceanography (studies of nutrient chemistry and suspended matter), and marine geology (studies of sediment, nodules, and acoustic stratigraphy). DOMES II involved monitoring the effects of the OMI and OMA pilot trials described below. The OMI and OMA systems required discharge into the sea of the return water from the lift system and this was characterized and monitored. DOMES was designed primarily as a data-gathering effort and final data reports were submitted to NOAA by early 1978. Data collected concurrently by other workers were included with the DOMES data for completeness, with a book of key findings and data published in 1979 [28]. The final official Draft Programmatic Environmental Impact Statement was issued in 1981 [29]. The conclusions of the study were: (i) The surface discharge from a typical riser lift system would have no long-term impact and would be very small at the scale of the CCZ, but a question remained on the impact of particulate matter on fish larvae, (ii) Impact from a typical seabed collector would be clearly adverse to benthic organisms at the site and recolonization rates were unknown, and (iii) The extent and impact of sediment plumes generated at the seafloor was still largely unknown and demanding of further investigation.

By the early 1980s the whole subject of the possible environmental effects of deep-sea mining was starting to be more fully investigated, and a complete issue of the journal *Marine Mining* (vol 3,1/2) was devoted to it in 1981. Initial concerns were focussed on limiting the disturbance as much as possible and devising remedial action, but gradually moved towards it becoming the wider issue of whether we should be disturbing possibly the last pristine environment left on Earth and therefore should we be undertaking deep-sea mining at all. However, such concerns were not fully articulated until well into the twenty-first century. The earlier view generally prevailed in the 1960s and 1970s, best summed up by Mero [10] who believed that nodules could be mined and processed with little or no effect on the marine environment because, being a new industry, it could be designed to be non-polluting from the start.

Law of the Sea and Deep-Sea Mining

The legal regime governing deep-sea mining is established by the 1982 United Nations Convention on the Law of the Sea (the Convention) and its 1994 Implementing Agreement (the IA). The history and implementation of this regime are complex and beyond the scope of the present book. Only major issues that influenced the course of seabed mining over the period covered in this book are briefly highlighted here.

In a speech before the UN General Assembly in 1967, Arvid Pardo, Ambassador for Malta to the UN, proposed the designation of the seabed as "a common heritage of mankind and should be used and exploited for peaceful purposes and for the exclusive benefit of mankind as a whole" https://www.un.org/depts/los/convention_ agreements/texts/pardo_ga1967.pdf. His proposal addressed a growing disquiet among less developed, landlocked, and geographically disadvantaged states, exacerbated by Mero's [1] optimistic assessment of their value, that marine mineral resources beyond national jurisdiction would only benefit developed states with the ability to exploit those resources (see, also Earney [7]).

Pardo's speech led to the UN convening a committee to study the implications of his proposals which led in 1969 to the UN General Assembly adopting a resolution calling for the possibility of a third Law of the Sea Conference, proposals for a seabed regime, and a moratorium on deep seabed exploitation until an international regime could be established. In 1970, the UN General Assembly called for the convening of a third Law of the Sea Conference (UNCLOS III). The 1982 Law of the Sea Convention was adopted in April 1982 after 11 sessions, representing over 90 weeks of work, by a vote of 130 States for, 4 against (including the USA, who requested the formal recorded vote), and 17 abstentions.

Archer [30] summarised the main provisions of the Law of the Sea Convention pertaining to deep-sea mining. The part of the seabed beyond national jurisdiction, was known as "the Area". The Area and its resources were described as being the common heritage of mankind and activities in the Area had to be carried out for the benefit of mankind as a whole. The States that are party to the Convention must administer the resources and control mining in the Area through an International Seabed Authority (ISA). Prospecting, exploration and actual mining in the Area would be conducted through a so-called parallel system, whereby the ISA (through an entity known as the Enterprise), independent States and private companies could do so, pursuant to a contract issued by the ISA after an application procedure. This included submission by the applicant of prospecting data for an area large enough to support two mining operations. Half the area would be taken by the ISA on behalf of the Enterprise, either to be mined by it directly, or to be mined on its behalf by operators licensed by it, and the other half would be mined by the applicant. Once a contract for activities to be conducted at a defined site had been awarded and production quotas agreed (see below), security of tenure was guaranteed as long as the contractual conditions governing those activities were met. Commercial operators had to transfer their deep-sea mining technology to the Enterprise at what at the time were deemed to be "fair and reasonable" prices, if the technology was not available on the open market. In addition, commercial operators had to transfer their technology to individual developing countries under certain circumstances. Furthermore, commercial operators had to assist the Enterprise in obtaining manganese nodule (and other mineral resources recovered from the Area) processing technology. In order to protect the interests of developing countries, production quotas could be established by the ISA. Provision was made for compensation of land-based producers if there were adverse effects on their economies that could be attributed to deep-sea mining. After 15 years from the start of the first commercial production, the system of exploration and exploitation would be subject to a Review Conference. Here changes in the arrangements made could be decided by a three-quarters majority of the Parties, but existing contracts could not be changed. Finally, a new and important aspect of the Law of the Sea Convention was the acceptance of Exclusive Economic Zones (EEZs) in which coastal states have sovereign rights for the purpose of exploring for and exploiting, conserving, and managing the natural living and non-living resources of the seabed and its subsoil. Under the Convention, certain regulations adopted by the ISA, particularly with regard to the protection of the marine environment, also govern mining activities.

The seabed mining provisions of the Convention did not attract universal support, especially among the developed countries. Production quotas to protect land producers and the mandatory transfer of technology were especially problematic. Some established producers of the metals, in particular from developing countries, feared economic difficulties if deep-sea mining went ahead. Potential deep-sea miners feared that production quotas would limit the profitability of their operations and even threaten their viability. Furthermore, States that were importers of the metals in question were also concerned about production quotas, fearing supply limits of crucial metals. Under the technology transfer provisions, if the Enterprise or entities licensed by it could not obtain the technology needed to mine nodules on "fair and reasonable terms and conditions," then a contractor must "at a fair market price" transfer to the Enterprise or its licensee the technology being used in the mining. This included equipment, manuals, and everything required to conduct an effective mining operation. The contractors were also required to train Enterprise and/or designated personnel in the use of their equipment. Although some of the equipment to be used in deep-sea mining was likely to be obtainable on the open market, there were proprietary items that contractors would not wish to be released, of which nodule collectors were probably the most commercially sensitive. The foregoing issues, as well as the requirement that the US must pay 25% of the costs of the proposed International Seabed Authority, were among the major reasons for the decision by the USA to vote against the Convention, President Regen stating that "the deep-sea mining part of the Convention does not meet US objectives".

According to Ryan [31], the decision of the USA to vote against the Law of the Sea Convention was broadly welcomed by US Industry, including the National Ocean Industries Association and the American Mining Congress, summarized as not providing US firms with access to seabed minerals under reasonable terms and conditions. Essentially, Ryan [31] concluded that the the Convention required the developed nations to provide the wherewithal to undertake deep-sea mining while leaving the ownership and control of the resources to others.

The Law of the Sea Convention was voted for by the vast majority of States, but several abstained, most notably the UK, USSR, and W. Germany. Commenting on the lack of universal signing, a US Govt official commented that there would be no seabed mining under the treaty and that any such mining would have to take place outside it [32]. The Economist [33] re-stated the oft-proposed outcome of failure to achieve a universally acceptable Convention, namely that the international capital needed to finance deep-sea mining would unlikely to be forthcoming.

To facilitate the eventual operation of the ISA, which could not be set up until after the Convention came into force, a Preparatory Commission (Prep Com) was established. One of its activities was to protect the interests of the then-existing entities interested in deep-sea mining. These were the so-called Pioneer Investors (PIs), entities that had commenced deep-sea mineral exploration before the Convention was adopted, namely the existing consortia, France, India, Japan, and the USSR. Many problems soon emerged, including overlapping claims. As each potential PI had acted independently during the pre-Law of the Sea Convention period, some had explored and subsequently wanted to mine in the same areas. This was understandable as the best nodule mining areas are not uniformly distributed throughout the CCZ (note that at the time, only nodules were the focus of Prep Com discussions, polymetallic sulphides, and Fe-Mn crusts as resources of the Area became of interest to the ISA only in the twenty-first century). Overlapping claims had to be resolved before licenses could be issued by the Prep Com. Ultimately, all overlapping claims, both among signatories and non-signatories to the Law of the Sea convention were resolved.

It became clear soon after the conclusion of UNCLOS III that the Convention was not going to provide the framework that commercial deep-sea miners needed to progress the deep-sea mining industry in the Area. Indeed, some potential miners had harbored grave doubts about it even before the Conference was concluded. According to Verlaan [34], issues of difficulty that arose after UNCLOS III was concluded were not only production limitations and transfer of technology, but also the Enterprise itself, costs to States Parties, the compensation fund, financial terms for commercial operators, decision-making, environmental considerations, and the Review Conference. State Enterprises were not as inconvenienced by the Convention as was private industry, perhaps because their needs and interests differed from those of the industrial consortia which had to make a profit in order to remain operational. Whether or not the Law of the Sea Convention in its original form would have sounded the death knell for deep-sea mining is now a moot point, as the adoption of the 1994 Implementing Agreement (see Chapter "Economic and Legal Developments, 1983–2000") resolved several contentious issues. However, its original mining provisions as set out in Part XI and the Annexes certainly didn't help.

References

1. Mero JL (1965) The mineral resources of the sea. Elsevier, London, 312pp
2. Mero JL (1959) The mining and processing of deep-sea Manganese nodules. Institute of Marine Resources, University of California, Berkeley, 96pp
3. Tooms JS (1967, November) Marine minerals in perspective. Hydrospace
4. Dunham KC (1969) Geology and the natural environment of man II: seas and oceans. Q J Geol Soc Lond 124:101–129

5. Hornig DG (1966) Effective use of the sea. Report of the panel on oceanography (President's Science Advisory Committee). US Govt Printing Office
6. Johnson LB (1967) Marine science affairs in a year of transition (Report of the President to Congress). US Govt Printing Office, Washington
7. Earney FCF (1990) Marine mineral resources. Routledge, London, 387pp
8. Mero JL (1972) Potential economic value of ocean-floor Manganese nodule deposits. In: Horn DR (ed) Papers from a conference on ferromanganese deposits on the ocean floor. Arden House, Lamont Doherty Geol. Obs, Columbia University, January 20–22. IDOE National Science Foundation, Washington DC, pp 191–203
9. Mero JL (1977) Economic aspects of nodule mining. In: Glasby GP (ed) Marine manganese deposits. Elsevier, London, pp 327–355
10. Mero JL (1978) Ocean mining: a historical perspective. Mar Min 1:243–255
11. Glasby GP (1986) Marine minerals in the Pacific. Oceanog Mar Biol Ann Rev 24:11–64
12. Gerrard R (1976) The inter-university manganese nodule project. In: Economic geology of the sea floor. Symposium 108.1, 25th International Geological Congress
13. Andrews J, Friedrich G et al (1984) The Hawaii Tahiti transect. Mar Geol 54:109–130
14. Glasby GP (2002) Deep seabed mining; failures and prospects. Mar Georesour. Geotechnol. 20:161–176
15. Cronan DS (2015) Deep sea minerals. Geoscientist 28 (8). Geol Soc Lond:10–15
16. Pendley WP (1982) The US will need seabed minerals. Oceanus 25(3):12–17
17. Medford RD (1969a) Marine mining in Britain I. Min Mag 12(5):369–381
18. Medford RD (1969b) Marine mining in Britain II. Min Mag 12(6):474–480
19. Drechsler HD (1972) The Manganese nodule industry: a first approximation. In: Horn DR (ed) Papers from a conference on Ferromanganese deposits on the ocean floor. Arden House, Lamont Doherty Geol. Obs, Columbia University, January 20–22. IDOE National Science Foundation, Washington DC, pp 1–7
20. Clark JP (1982) The rebuttal: the nodules are not essential. Oceanus 25(3):18–21
21. Meadows DH, Meadows DL, Randers J, Behrens WW (1972) A report for the club of Rome's project on the predicament of mankind. Earth Island, London
22. Broadus JM (1987) Seabed materials. Science 235:853–860
23. Fellerer R (1979) German activities in the field of nodules. In: Proc Germinal seminar on off-shore mineral resources, Orleans 1978 (Documents du BRGM No 7), pp 427–435
24. Wright R (1976) Ocean mining, an economic evaluation. Ocean mining administration report, US Department of the Interior, 18pp
25. Lavelle JW, Ozturgut E (1981) Dispersion of deep-sea mining particulates and their effect on light in ocean surface layers. Mar Min 3:185–212
26. Lipton I, Nimmo M, Parianos J (2016) TOML Clarion-Clipperton zone project. Pacific Ocean. NI43-101 report. AMC Consultants, Brisbane. www.sedar.com
27. Parianos J (2017) History of development of the Clarion- Clipperton Zone. https://www.lulu.com/shop/john-parianos/history-of-development-of-the-clarion-clipperton-zone/paperback/product-1vj9nm22.html?page=1&pageSize=4
28. Bischoff JL, Piper DZ (eds) (1979) Marine geology and oceanography of the Pacific Manganese Nodule Province. Plenum Press, 842pp
29. DOMES (Deep Ocean Mining Environmental Study) (1981, September) Final program-matic environmental impact statement. U.S. Department of Commerce, National Oceanic and Atmospheric Administration, Office of Ocean Minerals and Energy, 295 p
30. Archer AA (1983, March, 23–32) Resources policy
31. Ryan PR (1982) Editorial Comment. Oceanus 25(3):2
32. The Times. December 11th 1982
33. The Economist. May 8th 1982
34. Verlaan PA (2020) Future of deep sea mineral resources: environmental issues. In: Legal, scientific and economic aspects of deep seabed mining: the International Seabed Authority at 25. Brill, Leiden

Activities on Manganese Nodules During the Post-war Boom

Introduction

One of the earliest controversies concerning manganese nodules was the source of the metals that they contain. Murray in Murray and Renard [1] thought that they were derived from the alteration of submarine volcanic rocks, while his co-author favored the neptunist argument that they were formed by precipitation from seawater of elements that were ultimately derived from the continents. Other suggested sources have been submarine volcanic springs and diagenesis. Diagenetic sources of metals are those from within the sediment on which the nodules lie and come largely, but not exclusively, from the decay of organic material containing those metals, as opposed to hydrogenous sources which are from the overlying seawater (Renard's favoured source). However, as a result of studies on manganese nodules collected during the 1950s and 1960s [2–4] it became apparent that the metals in nodules do not accumulate just from any one source but that metals from any of the proposed sources are potential contributors to them. The issue thus became to determine which of the possible sources is most important in supplying metals to any particular suite of nodules. Progressive mapping and collection of nodules during the ensuing decades helped to accumulate sufficient data to at least partially solve this problem (see [5] for a review).

Pacific Ocean

One of the immediate effects of the realisation outlined in the previous Chapter that manganese nodules could be of economic importance was the need to increase knowledge on their worldwide distribution and compositional variability. Much of the initial work on this was done in the Pacific by ships of the Scripps Institution of

© The Author(s), under exclusive license to Springer Nature Switzerland AG 2024
D. S. Cronan, *Deep-Sea Minerals Developments in the 20th Century*,
https://doi.org/10.1007/978-3-031-52342-7_2

Oceanography, sometimes in conjunction with the monitoring of the US western Pacific nuclear bomb tests in the 1950s [6]. Additionally, Menard and Shipek [7] reported on manganese nodule investigations during the Scripps Institution of Oceanography Downwind Expedition (1957–1958) finding them to occur extensively in the areas that they traversed.

Of the early post-war manganese nodule workers, John Mero [8–10] was most influential in drawing attention to the importance of the deposits, although Russian workers also made considerable contributions in this regard [11]. Mero [10] concluded that apart from the continental margins and ocean trenches there appeared to be no extensive area in the Pacific between 50 N and 60 S where nodules did not occur. Mero noted that nodules were most abundant in the central region of the Pacific, which he believed to be the result of there being the lowest rates of accumulation of sediments associated with the nodules in that region. He estimated there to be 1.66x10000000000000 tonnes of nodules in the Pacific but noted that Zenkevitch and Skornyakova [11] estimated only 0.9x1000000000000 tonnes. This difference illustrated the amount of work that still needed to be done on nodule abundances in the Pacific before meaningful nodule resource estimates could be made.

In terms of Pacific nodule compositional variability, based on both old data and newly analysed nodules, Mero [10] divided the Pacific into four compositional regions, A, B, C and D. A regions were high in iron and generally lay along the continental margins although one in the southwestern Pacific Basin was noted. A region nodules averaged 28.3% Fe, 21.7% Mn, 0.35% Co, 0.46% Ni and 0.32% Cu. B regions were high in Mn and were found mainly near the west coasts of the Americas. They contained nodules with on average 49.8% Mn, 2.3% Fe, 0.055% Co, 0.26% Ni, and 0.14% Cu. C regions were high in Ni and Cu and found in those parts of the Pacific furthest from the continents and were thought to be the dominant compositional regions of the Pacific. Their nodules contained, on average, 33.3% Mn, 17.7% Fe, 1.52% Ni, 1.13% Cu and 0.39% Co. The D region contained nodules high in Co and was found in the south-central part of the Pacific. Nodules there contained, on average, 28.5% Mn, 22.6% Fe, 0.66% Ni, 0.21% Cu and ranged in Co content from 0.7% to 2.1%. (all values were on a detrital free basis.)

Following on from Mero's pioneering work, Cronan [4] presented almost 100 new Pacific nodule analyses. Combining these with the data in Mero [10] and all other available data,there were enough data points in the Pacific for an updated regional compositional variability map to be prepared [12] (Fig. 1). Manganese was found to be greater than 20% in the pelagic areas of low sedimentation rates in the northeast tropical and southeast Pacific and to decrease in a westerly direction. Iron behaved in the reverse manner to Mn and was generally low in the eastern Pacific with values averaging between 5% and 10%. It generally increased in a westerly direction reaching maximum values in some of the island groups of the western Pacific. In general, Ni and Cu followed Mn and were highest in the pelagic areas of the northeast tropical Pacific. Both elements decreased in a westerly direction and low values of each generally occurred in the island groups of the South Pacific. Cobalt varied inversely with Ni and Cu and was found to be highest among the island groups of the South Pacific. Other elements were found to show less distinct

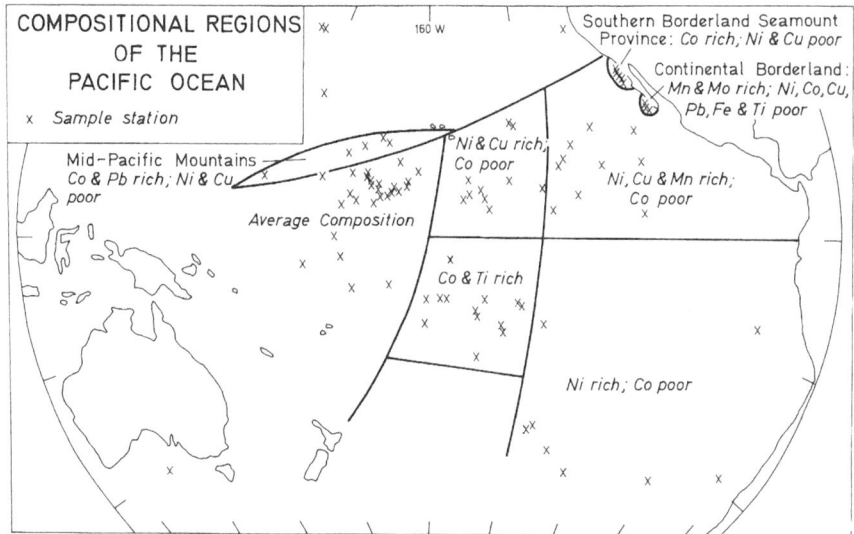

Fig. 1 Manganese nodule compositional zones in the Pacific Ocean, as perceived in 1972

variations, but nevertheless exhibited definite trends. Zinc and Mo were enriched in the pelagic areas of the northeast tropical Pacific, whereas Ti and V were more enriched south of the equator and especially in nodules from the island groups of the south and west Pacific. No nodules from the equatorial area were available for analysis and subsequently it was found that this is an area containing few if any nodules over most of its sea floor [13].

A summary of the extensive Russian work in the 1960s and 1970s on the zonal regularities in occurrence, morphology and chemistry of Pacific manganese nodules was given by Skornyakova [14]. Nodule abundance was found to be widely variable. Highest concentrations were confined to depths at and below the Calcium Carbonate Compensation Depth (CCD), the depth at which calcium carbonate dissolves in the oceans. Three latitudinal zones where nodule coverage of the Pacific sea floor was often greater than 25% were defined. The tropical and sub-tropical Pacific contained two of these zones, one on either side of the equator, at 7–30 N and at 8-40 S, and the third was in the sub-Antarctic region. Nodule coverage in the northern belt was commonly from 10% to 50% and averaged 7.8 kg per sq. m. Nodule coverage in the southern belt was greater, commonly more than 25%, with an average abundance of 18 kg per sq. m. Regional variations in nodule composition were found to be similar to those outlined by other workers. Manganese ranged from 5% to 34.5% but was commonly between 10% and 20% with higher values displaced towards the eastern Pacific. Iron ranged from 3% to 26% but was usually between 10% and 20%. High concentrations occurred on submarine elevations. The regional variations of Ni and Cu were found to be similar to that of Mn. Highest concentrations were found in the north-eastern tropical Pacific with a subsidiary area of Ni, but not so much of Cu, enrichment in the south-eastern Pacific. Cobalt ranged from less than 0.2% in the tropics to more than 1% in nodules from parts of the South Pacific. A strong depth dependency for Pb and V, as well as for Co, was noted, as were higher than average Co, Ti and Pb concentrations in South Pacific and Central Pacific Basin nodules.

According to Raab [15] work on the morphology and surface texture of Pacific nodules showed that there were two main textural types, those with smooth surfaces (s type) and those with rough or gritty surfaces (r type). Sometimes both textural types were found to occur in the same nodule. Where this occurred the smooth surface was the one exposed to seawater and the rough surface the one buried in the underlying sediment. A knobby equatorial band often coincided with the seawater / sediment boundary where the r and s type textures met. It was thought that the s type surface derived its metals from seawater and was thus hydrogenetic in origin and the r type surface derived its metals from the pore waters of the underlying sediments and thus was diagenetic in origin.

By the early 1970s the main features of the variability in Pacific nodule nature, distribution and composition were established. These would be refined considerably in subsequent decades but their main characteristics did not change. They showed that there were areas in both the northern and southern Pacific that offered promise for future manganese nodule mining. In the North Pacific, these were the Clarion-Clipperton Zone (CCZ) and the Central Pacific Basin, and in the South Pacific the Peru Basin and Penrhyn Basin (Figure 1 of "The Post-Second World War Deep-Sea Minerals Scene").

North Pacific Ocean

Clarion-Clipperton Zone (CCZ)

Of the nodule bearing areas of the North Pacific of potential economic interest, the CCZ between 6.5 N-20 N (Figure 1 of "The Post-Second World War Deep-Sea Minerals Scene") appeared to be the most promising. During the 1960s and 1970s it was mainly US concerns that worked there, but as mentioned in Chapter "The Post-Second World War Deep-Sea Minerals Scene", other countries such as France and Germany started to show an interest there too. These interests continued to develop after the US Consortia had ceased operations in the CCZ (see Part II) and, indeed, Germany still maintains an active interest in CCZ nodules at the time of writing. Horn et al. [16] found the CCZ to be the site of "widespread and intense development of manganese nodules". Their abundance and distribution were thought to be a function of a low rate of deposition of their associated sediments, which had rates of around 1–3.5 mm per 1000 year. Compositionally the deposits were found to be Mn, Ni and Cu rich.

Mero [17] provided some useful information on the early US industrial interest in CCZ nodules. He pointed out that INCO (International Nickel Company) had sponsored some investigations on nodules as early as 1959 and in 1974 decided to embark on a major development programme with the West German AMR Group and DOMCO, a group of Japanese companies. In 1961, Kennecott Copper became interested in nodules and late in 1973 went into association with Noranda, Mitsubishi, Rio Tinto Zinc, Consolidated Goldfields and BP. In 1962, the Newport News Shipbuilding and Drydock Company became interested in nodules and in 1964 together with Dow Chemical embarked on a cruise in the Pacific on the "Prospector" (the first vessel expressly fitted out for nodule exploration) but did not find any viable deposits of nodules in the areas studied. Lockheed became interested in nodules in about 1964. The Lockheed nodule mining vehicle was the only one of the early mining vehicles to be publicized [18]. In 1976, they went into association with Amoco Minerals, Billiton and Westminster Dredging. According to Mero [17], the major US nodule mining groups were expected to spend something in excess of 300 million dollars to complete their initial programmes.

The early work of the US Industrial Groups on manganese nodules has been described by Lipton et al. [19] and Parianos [20] based largely on participant sources quoted therein. The International Nickel Company (INCO) first became interested in deep-ocean mining for manganese nodules in 1958. During its early work, INCO contracted with others (John Mero, Dames and Moore, and Deepsea Ventures) to assist with exploration activities including several survey cruises. INCO then explored in their own right with the M/V Growler. As mentioned, Kennecott Copper Corporation became interested in manganese nodules in the early 1960s. Through their subsidiary Bear Creek Mining, Kennecott's first exploration cruise dates back to 1962, when 10 tons of nodules were dredged from a site west of Baja California. Kennecott followed this up with two more cruises in 1967, the "Clarion" cruise, and the "Confidence" cruise which investigated a portion of the CCZ. A subsequent "Aries" cruise in 1968 was a 3 month effort in the eastern and central tropical Pacific which resulted in delineation of what Kennecott called the Albatross Deposit. In 1969, Kennecott Exploration Inc. (KEI) was formed and between 1970 and 1974 KEI conducted another seven cruises, with progressively more detailed sampling and survey work. In 1970, on the Crux Cruise, KEI discovered what they called the Frigate Bird deposit in the eastern CCZ . The exploration work described above ended up with two areas of interest for KCON (i) The Albatross area in the central part of the CCZ, which ultimately became the "USA 4" license in 1984 (U.S. Department of Commerce) under the Reciprocating States Agreement and (ii) The Frigate Bird area in the eastern part of the CCZ which ultimately became the "UK" license in the early 1980s. Deepsea Ventures' initial work started in collaboration with the Newport News Shipbuilding and Drydock Company. Their first cruises were also in the mid-1960s. Deep-Sea Ventures, and associates, went on to focus their work in the central CCZ. Lockheed and associates conducted an extensive exploration campaign across the CCZ utilising the M/V Governor Ray between 1978 and 1981. Work included sampling, photography and video survey.

The involvement of the US companies in early manganese nodule exploration and mining development in the CCZ has also been reviewed by Wright [21], much of it obtained either directly from senior industry sources or derived from their appearances before Govt committees, or from industry statements. Likewise, Earney

[22] has given a comprehensive account of their later activities when the US companies entered into partnerships with various non-US companies to form the following Consortia:-

Ocean Management Inc. (OMI), Inco, Metallgessellschaft AG, Preussag AG,Sedco-Forex and DOMCO
Ocean Minerals Company (OMCO), Lockheed, Cyprus Minerals
Kennecott Consortium (KCON), Kennecott, Standard Oil, BP, Rio Tinto, Consolidated Goldfields, Noranda, Mitsubishi Corp
Ocean Mining Associates (OMA),Deep Sea Ventures, US Steel, Union Miniere, Ente Nazionale Idrocarburi

Wright [21] pointed out that the embryonic ocean mining industry had, by the mid-1970s, been carrying out research and development on deep ocean mining for over 10 years, almost all of it on manganese nodules in the CCZ. Each principal ocean mining company had identified nodule deposits that could provide mine sites and in the case of Deep Sea Ventures a claim had been made to a preferred deposit. To varying degrees, all the companies had designed and sometimes tested components of nodule mining systems. Nodule processing had been evaluated in most cases and some companies had completed successful tests of their preferred processing option.

According to Lipton et al. [19], French interest in the exploration and mining of Pacific nodules dates back to the mid-1960s but field work commenced in early 1970 when the Company "Société Le Nickel" (SLN) and the National Centre for the Exploitation of Oceans (CNEXO), which later became today's Ifremer (Institut Français de Recherche pour l'Exploitation de la Mer), began studying the nodule deposits, focusing not in the CCZ, but near the Marquesas Island Group in French Polynesia (South Pacific), but this exploration did not result in the finding of any deposits of likely commercial interest. Work in the CCZ followed soon thereafter with it being explored in 1975 and 1976. In 1974, the Association Française d'Étude et de Recherche des Nodules Océaniques (AFERNOD) was formed. AFERNOD focused on the CCZ and used four research ships for most of their work during the 1970s, Coriolis, Noroit, Suroit, and Jean Charcot. They also used one of the world's first AUVs, the Epaulard, and the manned submersible Nautile (see Part II).

The Japanese Government and companies were prolific participants in CCZ nodule exploration and development in the 1970s. Japanese government efforts started in 1974 using their newly commissioned research vessel the Hakurei-Maru. From 1975 through 1979 the Ocean Development Office of Japan's Ministry of International Trade and Industry (MITI) carried out cruises for about 90 days a year studying manganese nodules in the CCZ southeast of Hawaii using the 'Hakurei-Maru. The Metal Mining Agency of Japan oversaw the work. After 1980 the number of days increased and the vessel used was the more advanced Hakurei-Maru No 2 which entered into service in 1980.

Central Pacific Basin (CPB)

While always maintaining their main interest in the CCZ, the Japanese undertook a study of manganese nodules in the Central Pacific Basin (Figure 1 of "The Post-Second World War Deep-Sea Minerals Scene"). According to Usui and Moritani [23] the occurrence, morphology and composition of manganese nodule deposits in the CPB and adjacent areas were studied by the Geological Survey of Japan (GSJ) from 1974 to 1983 using data from about 1400 bottom samplings, 900 seabed photos, 1000 chemical analyses and 800 mineralogical analyses, together with relevant seismic profiling and seismic stratigraphy.

The CPB surveys demonstrated that there were regional distribution patterns there for the two main surface textural types of nodules (r type and s type). The s-type nodules of hydrogenous origin were found typically in the northern CPB (8–13 N) where they were associated with clay sediments. In contrast, r-type nodules of diagenetic origin were found to be distributed widely in the equatorial central to southern CPB, in low abundances associated with siliceous sediments. R-type nodules were always enriched in Mn, Ni and Cu,as they are in the CCZ, in contrast to s-type nodules that were found to be richer in Fe and Co.

South Pacific Ocean

Of major importance in initiating work on manganese nodules in the South Pacific was the New Zealand Oceanographic Institute (NZOI) which in the early 1970s put in place a program to map nodule distribution there. Glasby and Lawrence [24] prepared a chart series with all published nodule data from the South Pacific on it. From these charts, manganese nodule occurrences in the South Pacific were classified by Glasby [25] as being from (a) basin environments such as the southwestern Pacific Basin, Peru Basin, Chile Basin, South Tasman Basin, etc., (b) the circumpolar belt approximately 1000 km wide and (c) the seamount regions of the Manihiki Plateau, Society Islands, and the Tuamotu Archipelago. In addition, limited quantities of nodules were found in marginal plateau and seamount environments such as the Campbell Plateau off NZ and the Tasman Seamounts off Australia. Glasby [25] reported that nodules were largely absent from the East Pacific Rise. The NZOI also mounted two nodule exploration cruises in the South Pacific using the RV Tangaroa, one in 1974 to the Southwestern Pacific Basin and one in 1976 to the Samoan Basin (see below).

Peru Basin

Manganese nodules of possible commercial interest were discovered in the Peru Basin (Figure 1 of "The Post-Second World War Deep-Sea Minerals Scene") during the early phase of Pacific nodule investigations. Thijssen et al. [26] reviewed early work there. In his review of the literature on South Pacific nodules, Glasby [25] suggested that the Peru Basin was likely to be only of marginal interest from the economic point of view, although nodules with a combined Cu and Ni content of up to 2.54% were noted there. It was in June 1978 that the first detailed survey of manganese nodule occurrences was made in the northern part of the Peru Basin in depths close to the CCD using the then newly constructed state-of-the-art marine minerals exploration vessel R V Sonne [26]. The cruise (SO-4) was arranged by the West German AMR industrial group which had manganese nodule mine site interests there. Two further cruises to the Peru Basin by Sonne were carried out in 1979 [27]. During cruise SO-4, free fall grab samples were taken from four areas in the Peru Basin's northern sector. Of note was the substantial variability in nodule size and density distribution. The Sonne was one of the first research ships to have onboard analytical facilities and these were used to good effect to analyze the recovered nodules for Mn, Fe, Cu, Co and Ni. They had high Mn/Fe ratios, intermediate Cu and Ni, and low Co. Analytical results are given in [26] and the results of more detailed investigations in [28]. During cruise SO-13 in the Peru Basin in 1980 further investigations were carried out to assess the economic potential of nodules there A detailed account of the whole Peru Basin nodule program has been given by von Stackelberg [29].

Southwestern Pacific

Some of the oceanwide nodule studies of the 1960s and early 1970s outlined above pointed to parts of the southwestern Pacific as having a nodule mining potential. In particular, the work of Skornyakova and Andrushchenko [30] allowed attention to be focussed on these areas. The tropical and subtropical islands of the southwestern Pacific are mostly poor in resources, and as their land areas are minuscule compared with the sea areas around them, (over which they soon were to have legitimate EEZ claims under the Law of the Sea Convention), it was natural for them to look offshore in their search for mineral resources. Glasby et al. [31] believed that one reason for the delay in their being considered was their remoteness from the main industrial centers of the northern hemisphere. To this could be added the belief that as the CCZ appeared to offer so much promise, there was little need to look elsewhere for nodules at that time. Nevertheless, there was a growing interest during the 1970s by the mostly newly independent island nations of the southwestern Pacific in understanding the nature of the resources in the seas around them. The setting up of the Committee for Co-ordination of Joint Prospecting for Mineral Resources in South Pacific Offshore Areas (CCOP/SOPAC) in 1972 was pivotal in this regard. It rapidly established itself as an important resource-oriented force in the region and its work included a number of cruises to investigate nodule distribution around its member countries [31].

The burgeoning importance of offshore resources in the consciousness of the peoples of the southwestern Pacific island nations is well illustrated by the considerable interest that was generated in the Cook Islands after the cruise in their waters by the NZ research vessel Tangaroa which in 1974 discovered substantial amounts of manganese nodules there. According to Kingan [32], although the discovery was of significance, its importance was overplayed to the then Cook Islands prime minister, giving him the impression that these nodules were a source of immediate wealth to the Cook Islanders. Around this time there was an election campaign in the Cook Islands, during which the value of the nodules to the Cook Islanders was trumpeted by the party in power. On the eve of the election, the prime minister went on the radio with a special news report saying falsely that the leader of the opposition had agreed to sell the Cook Islands nodules to Howard Hughes, the US industrialist, who at that time was believed to be in the nodule mining business (see below). The announcement was made too late for the opposition to respond and Kingan believed that it was this false announcement that lost them the election. Later, a technical report of the work of the Tangaroa on its 1974 SW Pacific Basin cruise was given by Glasby et al. [33]. The RV Tangaroa did discover a large province of nodules in the Southwestern Pacific Basin south and southwest of Rarotonga but they were found later not to be of any economic value. Subsequent early work on southwestern Pacific nodules has been reviewed by Exon [34] and Glasby et al. [31]. These studies demonstrated that by the end of the 1970s, it was evident that the then most prospective regions for potentially economically valuable nodules in the region lay in the deep ocean basins between the equator and 20 S. It was these areas that subsequently received the greatest attention for nodule exploration.

In 1975, another of the great marine mineral conferences that did so much to propagate early knowledge of the deposits and move work on them forward was held in Suva, Fiji, under the auspices of CCOP/SOPAC. Its proceedings were published [35] and represented the first attempt to gather SW Pacific marine minerals work into one volume. It was followed by the Fourth Session of CCOP/SOPAC, a largely administrative and policy-making meeting, at which a number of proposed projects dealing with marine resources, including manganese nodules, were discussed and recommendations made to advance studies on them [36]. This underpinned much of the deep-sea minerals work in the southwestern Pacific for the next decade (see Part II).

Transects

Following the initial oceanwide syntheses of manganese nodules data in the 1960s and early 1970s, more detailed studies focussing on transects from one nodule type region to another in the Pacific Ocean commenced in the late 1970s and continued into the 1980s. An important aim of this work was to investigate how nodules varied in abundance and grade between different depositional environments, and what caused the variations. One hoped-for benefit of this work was that if it was known which environments hosted the most potentially economic nodules, prospecting activities could be concentrated in areas where those environments occurred, thereby cutting exploration costs.

Hawaii-Tahiti Transect

The first of the transects was the Hawaii -Tahiti Transect, not directly between the islands but east from Hawaii to 134 W, and then south along the 134 W meridian, ending in Tahiti (Fig. 2). This was primarily a Franco-German-US study under the auspices of ICIME with other nations involved to a lesser extent, and using two

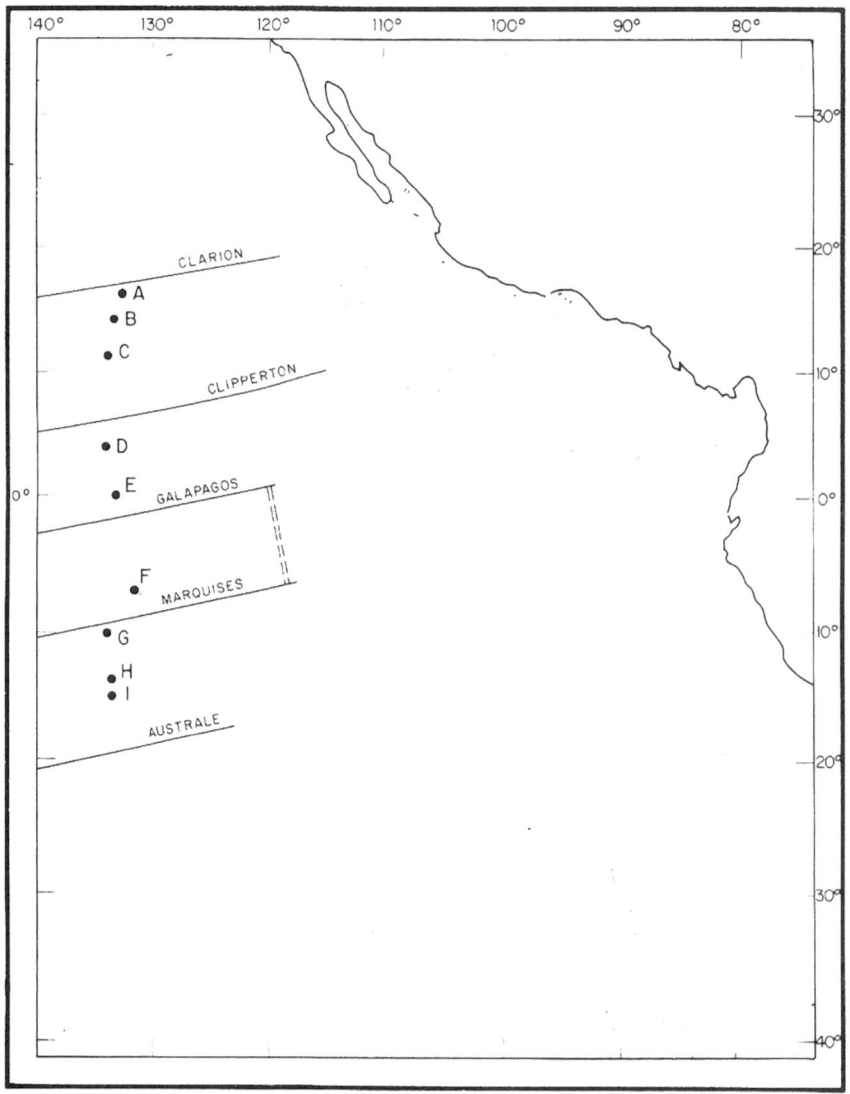

Fig. 2 The Hawaii-Tahiti Transect, 1978–1980 and detailed study areas on it occupied by RV Suroit and RV Sonne

ships, the Sonne and the Suroit. It was mainly carried out in the late 1970s [13]. Nine sites between 20 N and 20 S were studied in detail to examine both regional and local environmental associations of nodules. Variations in depth, microtopography and bottom currents, all superimposed on regional north-south variations in biological productivity, were found to be the main controls on nodule variability along the transect. There was a quasi-symmetrical distribution of nodules about the equator with nodules in the CCZ north of the equator, where strong Antarctic Bottom Water (AABW) flow was present, being large and regular in shape, whereas south of the equator where AABW flow could not be established the nodules were smaller, more irregular, and sometimes partially buried. Compositionally the nodules also varied about the equator. Manganese, Cu and Ni all increased towards the equator from the north and, more irregularly, from the south, while Fe behaved in the opposite manner. In the equatorial region itself, no nodules were found.

Wake-Tahiti Transect

The second major transect was the Wake-Tahiti Transect [37], this time directly between the two islands and traversing the Mid-Pacific Mountains (MPM), the CPB, the Manihiki Plateau and into the Penrhyn Basin. (Fig. 3). It was carried out from January to March 1980 by the Geological Survey of Japan using the Hakurei-Maru. Like on the Hawaii-Tahiti Transect, a quasi-symmetrical nodule distribution was found from north to south. In the MPM and northern CPB irregular to discoidal smooth surfaced (s type) nodules were found in which Mn, Cu and Ni were low. In the central to southern CPB, spherical rough surfaced nodules (r type) predominated in which Mn, Cu and Ni were higher. No nodules were found on the Manihiki Plateau. In the Penrhyn Basin, spherical to discoidal s type nodules were found, also with lower Mn, Ni and Cu. These regional distribution patterns were considered to support a close relationship of the nodules to their depositional environments. Environmental factors thought to be affecting the nature of the nodules included the rate of accumulation of their associated sediments, the nature of these sediments, breaks in sedimentation, bottom water flow, and surface biological productivity.

Aitutaki-Jarvis Transect

The third Pacific transect designed to investigate manganese nodule variability in relationship to environments of deposition was the Aitutaki-Jarvis transect, organized by Imperial College London. Again, it was a direct line between the two islands (Fig. 4) and was planned just after the earlier two transects in order to follow

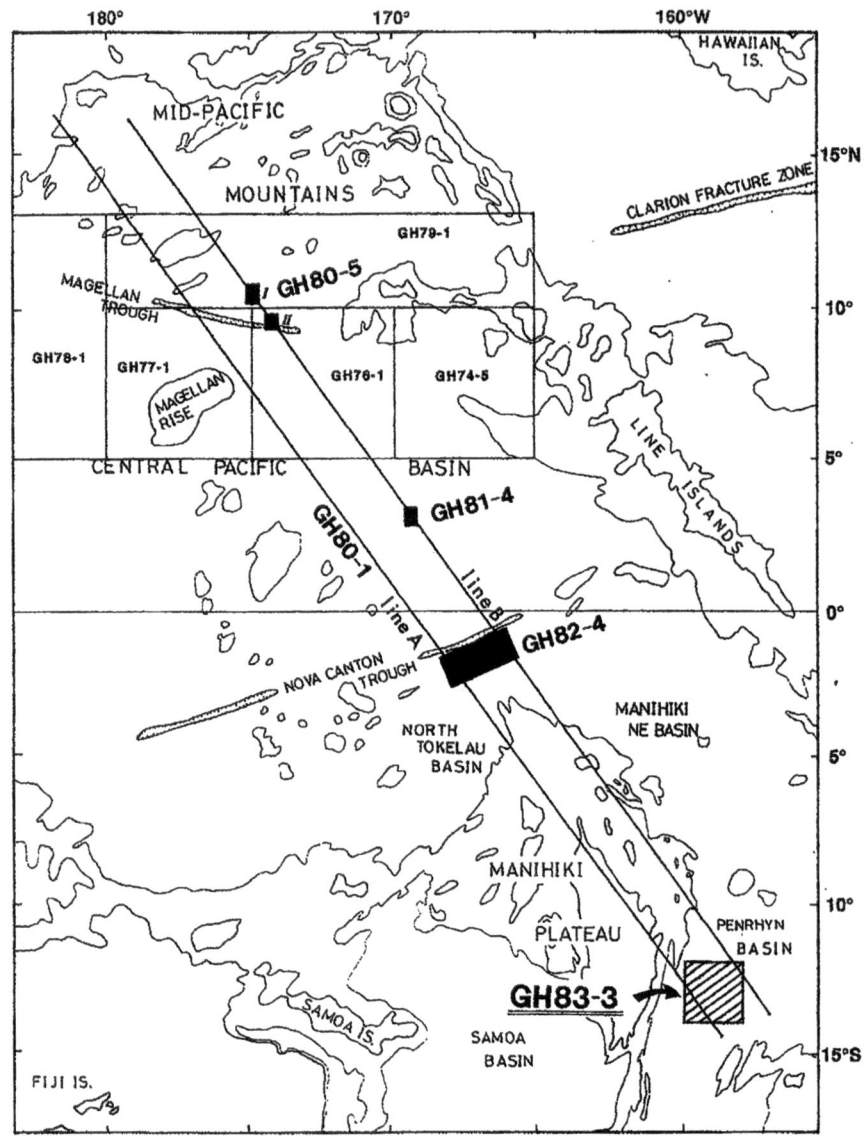

Fig. 3 The Wake-Tahiti Transect (1980) and detailed study areas on it as compiled by the Geological Survey of Japan in the late 1980s

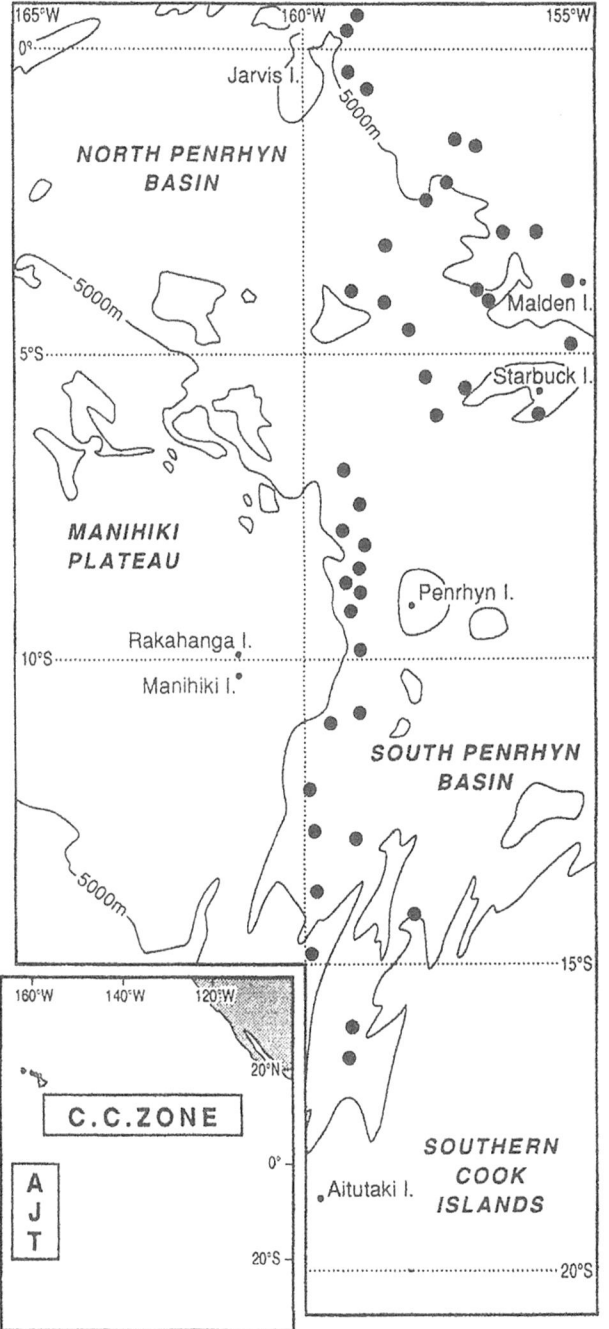

Fig. 4 The Aitutaki-Jarvis Transect (1987) and station groups occupied on it by the RV Thomas Washington

up on them.[1] The transect was N-S at 160 W between Aitutaki and Jarvis Islands, wholly in the equatorial and South Pacific as the tropical North Pacific at that longitude had been well explored by the Industrial Consortia. It found the same environmental controls on nodule variability that had been found on the two earlier transects but also found additional controls on nodule compositional variability such as the nodule's relationship to the CCD and the composition of the sediments associated with them [38]. Fluxes of Mn, Ni and Cu to the nodules were found to increase in sub-equatorial regions of high biological productivity but high values of these metals only occurred in nodules within about 200 m of the CCD.

Atlantic Ocean

Initial work on nodule distribution in the Atlantic Ocean [10] found that nodules were more limited there than in the Pacific Ocean, most likely as a result of the differing patterns of sedimentation in the two oceans. Sedimentation rates were thought to be too high to permit the development of extensive nodule fields. Much of the marginal Atlantic is floored by terrigenous sediments derived from the adjacent land areas and apart from a few areas where sedimentation is more limited was found to contain few nodules. The central Atlantic Ocean contains the Mid-Atlantic Ridge (MAR) a major volcanic feature, where again nodules are uncommon. It was concluded that it was only in the basins between the Mid-Atlantic Ridge and the terrigenous sediment floored areas adjacent to the continents that significant nodule deposits could be expected [39].

Regional variations in the composition of Atlantic manganese nodules [40] were found to be less marked than in the Pacific. Manganese was found to be greater than 15% in the basins on either side of the MAR, with the highest values of all occurring off South Africa. By contrast, it was found to be low in the South Atlantic, often less than 10% in the Drake Passage and Scotia Sea. Iron generally behaved the reverse of manganese. Nickel and Cu were generally low and found to follow Mn. Other than possibly on the Blake Plateau, Atlantic nodules have not been thought of as having any economic significance.

Mero [10] described nodules on the Blake Plateau as "an interesting deposit". The Blake Plateau is about 200,000 sq. km in extent off the SE United States with a depth of between 200 m and 1000 m. Sea floor photographs suggested a fairly heavy concentration of nodules there. Ferromanganese oxide pavements were also observed. Mero thought that the high nodule concentrations and the presence of sediment-free pavements might be due to the Gulf Stream, which crosses the plateau, sweeping it clean of terrigenous sediment. Analyses of four Blake Plateau nodules gave averages of 13% Mn, 13.5% Fe, 0.34% Ni, 0.12% Cu and 0.41% Co [10]. Fields of nodules and ferromanganese oxide slabs in the southeastern part of

---

[1] Because of the non-availability of a ship it was not carried out until 1987. The UK rarely had a research vessel in the Pacific in the 1970s and 1980s and it was not until NERC (the UK research ship agency) could arrange a ship swap with the NSF (the US research ship agency) under an arrangement that they negotiated in the mid-80s, that a ship became available for the work. It was carried out using the Scripps vessel Thomas Washington.

the area at water depths of around 800 m were investigated by Deep Sea Ventures in 1969–1970 using underwater TV and dredges and an area of about 40 sq. miles was designated to test a prototype airlift system for raising nodules.

As direct objects of mining, the Blake Plateau nodules and pavements were considered submarginal until the late 1970s when lease requests were first sought from the US Dept of the Interior (DOI) by Reynolds Metals International. In January 1982, the DOI announced a plan to open the US continental shelf to leasing for minerals. The Blake Plateau was an area chosen for priority leasing attention. It was felt that because the Blake Plateau deposits probably represented the largest domestic source of Mn and Co within US territorial jurisdiction, they warranted serious economic consideration even though their content of these metals was less than in prime Pacific nodules. This latter concern was partly compensated for by the proximity of the deposits to the US mainland and shallow depth of occurrence thought to facilitate easier mining. Additionally, the deposits were thought to have the potential for providing catalyst material for removing S, V, and Ni from petroleum feedstocks. The catalyst could subsequently be reprocessed to recover these elements. However, no mining has ever taken place on the Blake Plateau.

Indian Ocean

Mero [10] had very little manganese nodule data available to him from the Indian Ocean. More was supplied by Cronan [4]. Results are summarised in Fig 5. Additional data on Indian Ocean nodules was summarised by Bezrukov and

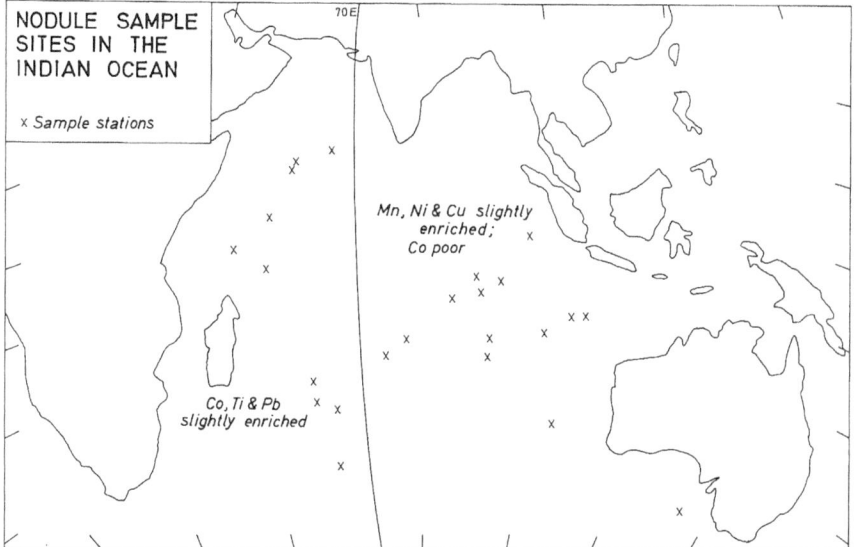

Fig. 5 Salient chemical characteristics of Indian Ocean manganese nodules as perceived in 1972

Andrushchenko [41] who found them to be more common in the basins than in the elevated areas, with highest concentrations occurring in the basins on either side of the 90 E Ridge.

The first wholly marine mineral exploration cruise in the Indian Ocean was by the German RV Valdivia (VA-07) in 1973–1974. The second leg of this cruise was from Mauritius to Singapore, on which manganese nodules were investigated by the German company Preussag A G, mainly in the Central Indian Ocean Basin (CIOB) where previous compilations of published data had indicated that the most potentially economic nodules in the Indian Ocean might occur. A British cruise took place on the RRS Shackleton in the NW Indian Ocean between Bombay and the Seychelles in 1975. Its track was designed to obtain nodules and/or crusts from the four main physiographic features in the NWIO, the Arabian Abyssal Plain, the Carlsberg Ridge (revisiting Area 4c of a 1963 RRS Discovery cruise carried out as part of the International Indian Ocean Expedition) where interesting manganese nodules had been found previously [42],the Somali Basin and the Seychelles Plateau. Results of the cruise were given at scientific meetings soon after [43, 44]. Direct Indian involvement in deep-sea minerals work at sea commenced with a cruise of RV Gaveshani in 1981 which sampled nodules on the Carlsberg Ridge [45]. Prior to this, the Indian National Institute of Oceanography had carried out oceanwide compilations based on published data which confirmed the Central Indian Ocean Basin (CIOB) as the area most likely to contain ore-grade nodules [46].

The early studies at sea on Indian Ocean nodules were supplemented by a compilation exercise carried out on existing Indian Ocean nodule data at the SIO by Frazer and Wilson [47], and additional new analyses made by Moorby [48] which together with all previously published data were used to tabulate and map the compositional variability in nodules throughout the Indian Ocean [49]. The physiographic complexity of the Indian Ocean was found to imprint itself on the distribution and compositional variability of the nodules that it contains. Nodules there were found to be sharply differentiated on the basis of physiographic features, of which only the Central Indian Ocean Basin was confirmed to offer any prospect of containing potentially economic nodules. At their highest, Cu, Ni, and Mn values in CIOB nodules were found to be comparable with those in some Pacific CCZ nodules. This area subsequently became the focus of a Govt of India program on Indian Ocean nodules that commenced in the early-1980s (see Part II).

Resource Potential

From the very earliest commercial interest in nodules, there was a need to quantify their resource potential. In order to do this, published data were compiled into databases, of which the SIO manganese nodule database was perhaps the best known. However, in such compilations, there was always the question as to how much confidence could be put in the chemical analyses available. There was the problem in ocean-wide compilations that most of the data had come from disparate sources, had been obtained in different laboratories by varying analytical methods, and presented in different ways. This exercised early workers on the resource potential of nodules using published analyses as to whether or not the data were representative enough to allow meaningful resource calculations to be made. This would have been less of a problem with nodule data collected by the Industrial Consortia because the techniques they used were more uniform, but these data were not then publically available, nor for the most part are they available at the time of writing. Frazer [50, 51] considered the issue of the value of publically available nodule data in resource calculations and concluded that because these data were likely to be distributed normally around the correct value, any erroneous values were likely to average out. Also, using data from several independent sources was thought likely to reduce the risk of systematic error such that any errors were likely to be random. Frazer pointed out that using a large database assists in reducing the effects of error. Even if errors of up to 10% are present in individual analyses (not uncommon in the 1950s and 1960s when the precision and accuracy of the analytical techniques available were poor by modern standards), these become random when analyses from several sources are added together. Frazer concluded that the use of nodule databases such as were available in the 1960s and 1970s was defensible for the purpose of the assessment of the regional distribution of elements in nodules and their resource assessment over large areas, helped by the fact that the values used were often averages of several samples from a single site. However, she advised that where single analyses were being considered, caution in the interpretation of the data was required.

One question that was often posed in the 1960s and 1970s was what actually constituted a manganese nodule reserve or resource. This subject was extensively investigated by Alan Archer of the UK Geological Survey [52–54]. To summarise, Archer [54] drew attention to the misuse of terms like reserve and resource when used in an economic geology context, pointing out that the layman would draw little distinction between these terms. According to Archer [54], the need to codify the terminology used in mineral evaluation at large had long been recognized by the USGS and US Bureau of Mines. What constituted mineral reserves and mineral resources in the wider context had been discussed by Blondel and Lasky [55] and included the definition of reserves as mineral material considered as being exploitable under existing conditions, and resources as the reserves and that mineral material which to be exploited needed more favorable conditions than those existing at the time of the assessment. However, according to Archer [54] this definition of a resource had a serious flaw because it included a sub-economic category, namely material that either requires a higher price or a major cost-reducing advance in technology to be economically viable. Thus he considered that one limit of what the resources of any mineral are is left to subjective judgment. Archer (1987) attempted to eliminate the confusion around what constituted marine mineral reserves and resources by suggesting that marine mineral resources should be defined as mineral occurrences on the seafloor that are likely to become workable in the next 20–30 years. On this basis no deep-sea mineral deposits were resources at the time as none were mined in the 30 years following Archer's work. Resources may become reserves with changing economic and technical circumstances.[2]

Mero (1965) considered that manganese nodules contained sufficient metals to last for thousands of years at then rates of consumption. However, other estimates differed. For example, the French Group [56] calculated that potentially recoverable nodules in the part of the North Pacific where they were working contained only 100 times the then annual world consumption of Ni and 8 times that of Cu. Also, there was no universal agreement at the time as to what nodule grade and abundance were actually necessary for a minable deposit. Most workers took the minimum abundance to be 10 kg per sq. m but Bastien – Thiry et al. [56] considered this value to be too high. Holser [57] also suggested an average of 10 kg per sq. m. Similarly, there was no firm agreement on what should constitute the average and cut-off grades for Ni and Cu needed to support a mining operation. Nevertheless, the average and cut-off contents of 2.27% and 1.8% combined Ni and Cu suggested by Kildow et al. [58] gained general acceptance.

How many manganese nodule mine sites might be available when mining actually started was also a subject of debate. Holser [57] attempted to quantify manganese nodule mine site availability. He defined a mine site as an area commercially exploitable for a single contract of 3 million tonnes of nodules per year for a period of 25 years. However, Glasby [59] questioned the possibility of achieving a 3 million tonnes per year throughput on the basis of technical considerations. Nevertheless, Holser's definition gained widespread acceptance and on this basis he concluded that between 190 and 460 mine sites, each yielding 75 million tonnes of nodules would be available on the basis of the advanced technology expected by the time that a significant number of projects were in operation [57]. However, he believed that relatively inefficient operations in the early phase of ocean mining would cause an *apparent (Holser's italics)* reduction in this number of mine sites to about 80–180. This was in general agreement with an estimate of 44–106 first-generation mine sites by Archer [52]. From estimates such as these, Archer concluded that the amounts of Ni and Cu that would become available from manganese nodule mining were unlikely to be either enormously greater nor enormously less than remain to be mined on land.

Economic Evaluations

As mentioned in Chapter "The Post-Second World War Deep-Sea Minerals Scene", in 1973 an oil crisis resulting in a sharp increase in oil prices had a dampening effect on progress towards deep-sea mining. Most aspects of nodule mining were expected

---

[2] In an attempt to simplify what are understood to be reserves and resources, Prof Alex Smith of London University suggested (pers comm circa 1989) that a simple analogy can be drawn with a bottle of whisky. If in a cupboard at home and available to drink at any time, the whisky is a reserve. If it is in a shop or bar down the road that are not always open, then the whisky constitutes a resource. If it is bought and taken home it becomes a reserve

to be energy intensive, but the proposed metallurgical processing of nodules in order to extract metals of interest was expected to be particularly energy intensive. The effect of the oil price increase on energy costs seriously impacted the perceived cost and profitability of ocean mining and prompted a reappraisal of the operation. Some authorities put processing costs at more than 50% of the total cost of manganese nodule exploitation. Coupled with increased processing costs, a second dampener on manganese nodule mining in the mid-1970s were uncertainties regarding the legal regime under which the nodules might be mined. As mentioned in Chapter "The Post-Second World War Deep-Sea Minerals Scene", the hoped-for resolution of this issue at the session of the Law of the Sea Conference in 1973 did not happen. Thus, by the mid-1970s there commenced a period of relative inactivity in the development of deep-sea mining. To reverse this, the legal issues needed to be resolved, and either energy prices had to come down so that nodule processing costs could be reduced or the prices of the metals to be recovered had to increase sufficiently to outweigh the increased energy costs.

The mid-1970s slowdown in nodule mining development permitted more rigorous appraisals of the economics of the endeavor than had been carried out hitherto. Notwithstanding uncertainties regarding the actual amounts of metals available for mining in nodules and the outlook for mining them, by the mid-1970s, detailed work on the economics of possible manganese nodule mining operations was being put into the public domain. One of the most comprehensive of these was released in May 1976 by the Ocean Mining Administration of the US Dept of the Interior [21]. The main objective of this study was to determine on a preliminary basis the economic feasibility of deep ocean mining operations. The basic approach was an analysis of the probable investment expectations which could be attributed to manganese nodule mining ventures being contemplated at that time, supplemented by a comparison of ocean mining economics with those of relevant alternative mineral extraction projects. The material for this analysis lay only to a limited extent in publically available information, so the major inputs were obtained by the Ocean Mining Administration from interview with Industry sources.

According to Wright [21] it was widely accepted at the time that deep ocean mining should be viewed mainly as a nickel recovery operation. Although the profitability of ocean mining ventures would also depend on the recovery of Co, Cu and perhaps Mn, nickel sales would yield almost 70% of gross revenues for most operations contemplated at the time. In 1976, therefore, perceived growth in the ocean mining industry in its early years would be competing with other sources of Ni. In the 1970s, Ni resources were generally classified either as sulfide or nickel laterite deposits. Most of the world's Ni bearing sulfide deposits were located in Canada, the Soviet Union, South Africa, and Australia. Nickel laterites were located primarily in tropical and subtropical areas. World demand for Ni averaged around 790,000 tonnes in 1973. World mine capacity was around 856,000 tonnes of which approximately one-third was comprised of nickel laterite deposits. Seabed Ni production had to be competitive with this land-based production if it were to gain a foothold in world markets.

Clearly, on the basis of the above figures, there was little need at the time to enter into ocean mining for Ni, unless it could be recovered from the deep sea at substantially less cost than on land. However, the Ni market was projected to expand at rates estimated from 2.6% to 7% annually. The significance for ocean mining of the projected increase in world Ni production was related to the comparative cost of mining Ni laterites. Thus, a measure of the likely profitability of ocean mining for Ni was a comparison with the projected returns from new Ni laterite mines. In other words, ocean mining economics should be compared with Ni laterite mining economics. Assuming an ocean mining operation of 3 million tonnes of nodules per year, Wright [21] concluded on the basis of her analysis that investment opportunities in ocean mining projects for Ni could be greater than in Ni laterite mining projects and that ocean mining for Ni could be competitive with land-based Ni ores for a proportion of the expected increase in world Ni production.

Wright [21] also considered the effect on manganese nodule economics of extracting manganese from the nodules. In the 1970s, a three-metal operation, Ni, Cu and Co,seemed to be the most likely contender for first-generation ocean mining. An exception was the Deep Sea Ventures consortium (OMA) which included a steel company, US Steel, which was interested in also extracting Mn. By far the greatest use of Mn at that time was in steel production. Additionally, "private sources" told the Ocean Mining Administration that processing of manganese nodule tailings for Mn was technically feasible and could yield around one million tonnes per year of Mn. Wright concluded that the possible adoption of a four-metal process recovering Mn as well as Ni, Cu and Co, would be largely dependent on the extent to which investment returns for the three-metal operation plus extraction of Mn from tailings compared with those for the four-metal operation.

A later economic evaluation of manganese nodule mining was carried out by the Massachusetts Institute of Technology (MIT) [60] and reviewed by United Nations [61]. It dealt with the anticipated costs of deep-sea mining from the R&D and prospecting stages to the processing of nodules and waste disposal stages. The cost estimates were based mostly on data gathered in 1975–1976. Among other things, the MIT cost model was designed to calculate the internal rate of return (IRR) and payback period assuming per pound prices of $4 for Co, $0.71 for Cu, and $2 for Ni. The MIT study addressed three main categories of issues (i) technological and financial prospects for exploitation; the project was seen as a risky venture with uncertainty stemming from several sources such as ore composition, mineral prices, recovery rate, and processing efficiency (ii) US Legislation relating to key issues including the right to mine, environmental regulations, tax depletion allowances and political risk coverage and (iii) international arrangements such as payment of royalties, duration of mining rights, production limits and technology transfer issues. The study presented detailed capital and operating cost estimates for a five-phase mining equation: prospecting, exploration, mining, transportation and processing. The components of operating costs were energy, labor, materials, fixed and miscellaneous costs, and costs associated with research and development. The mining system in the MIT study was a hydraulic lift system (see Chapter "Exploration and Mining Development") based on and controlled from a floating platform above the mine site. A rapid transport system was assumed for transferring the nodules from the mine site to the transport vessel and from there to the port and processing plant and ended with the disposal of tailings. The estimated costs depended largely on the type of technology used in the various stages of the operation. The revenue component was straightforward. Gross revenues were to be the receipts from metals sales using long-run average market prices in constant dollars. It was assumed that commercial operations would commence in the sixth year of a project, after 5 years of R & D, prospecting and exploration. The IRR would be 18.14% and the payback period would be 5.4 years. These figures were sensitive to changes in assumptions and initial conditions such as (a) changes in metal prices, particularly of Ni which accounted for two thirds of the revenue in the cost model (b) changes in production rate (c) changes in the assumed ore grade (d) changes in operating costs including labour costs (foreign labour, ie non US, was assumed) and (e) time delays. As would be expected, increases in metal prices would be beneficial to the IRR, as would increases in production rates and increases in the grade of the ore. Deleterious would be time delays such as might occur due to exceptional adverse weather conditions, labor disputes, and the substitution of US labour for foreign labour. Overall, the MIT study concluded that deep-sea mining would be profitable, but did not consider possible impacts of it on the economies of developing countries already producing Ni, Cu and Co at the time, and how they might react.

One of the main economic concerns about possible manganese nodule mining was the effect it would have on land producers of the metals concerned. Levy and Odunton [62] believed that earlier uncertainties surrounding this issue had been at least partially resolved by the legal protection offered by the Law of the Sea Convention. A limitation of production of manganese nodules was going to be an important method to be used to prevent disruption of the world metal markets by deep-sea mining carried out by those entities doing it under Law of the Sea Convention rules. Production limitations were to be based on nickel production. However, had mining started, because of differing concentrations of metals in the nodules and differing world production rates of them, the main market that would have been adversely affected by nodule mining would have been that of Co. There could have been reductions in the export earnings of Co exporting countries like Zaire, Zambia, Cuba, the Philippines, and Morocco. If Mn were to be recovered from the nodules, the next market to be affected would have been that of Mn. Gabon, Brazil, India, and Ghana could have been affected. Land-based Ni producers would not have been greatly affected because, as mentioned, the nodule production limitations envisaged under Law of the Sea Convention rules were based on Ni production. Finally, Cu producers would have had little to fear from deep-sea mining as deep-sea Cu production was not expected to exceed 1% of world Cu production even under the most optimistic nodule mining scenario envisaged.

In the early 1980s Nyhart and others at MIT [63] carried out a sequel and revision of the 1978 model, which was also reviewed by United Nations [61]. In this study, the project was divided into ten components, prospecting, R&D, mining, marine transportation, ore discharge terminal operations, onshore transportation, processing, waste disposal, marine support, general administration, and continuing preparations. The capital and annual operating costs were estimated, as well as the activities, facilities, and equipment required by these ten components. The results of the study showed that the IRR was 9.21% for the baseline assumptions, 21.96 for the most favorable set of assumptions, and −6% for the least favorable set of assumptions. The same UN review analyzed a 1982 economic analysis by J E Filipse of a pioneer deep ocean mining venture which was carried out to complement Nyhart's continuing MIT cost model studies. The results demonstrated among other things that control on operating costs was more important than capital costs at low rates of return, that three-quarters of all possible savings from design improvements were in the processing operation, and that the economics and efficiency of power and fuel should be major targets for system improvement as almost half of the base case operating costs for processing were for power and a quarter for fuel and water.

Based on the above, an optimal case was constructed composed of the best combination of factors producing high returns. The optimal combination included a three metal processing plant located in a port in the US Pacific Northwest, transport and mining ships built by non-US organisations, a 4-year operational schedule, no taxes, interest-free loans, and all preparatory R&D subsidized by the US Government. Much of this was highly unlikely. A more realistic combination was to assume 10% interest on debt, 30+% tax, and little or no subsidy. On the basis of such a scenario, the IRR ranged from 4% to 42.3% after tax [61].

Flipse [64] reviewed the various cost models. Assuming a 3 million tonnes per year three metal operation (Ni, Cu and Co), a basic return on investment "pay-out" model was constructed yielding an IRR of 7%. Filipse considered it unlikely that nodule mining would be undertaken on this basis unless "a critical feedstock for a company's major product or a national need for a strategic metal develops". He also noted that a four-metal operation would return little more to the investor than a three-metal project because the extra cost of processing in the former would offset the increased revenue from the recovery of manganese.

From the analyses outlined above it was thought in the early 1980s that manganese nodule mining could be profitable based largely on Ni recovery, but that the recovery of other metals would also be of significance. Of these, Co was of the greatest importance, but because the world market for it was much smaller than that for Ni, the effect of marine Co on the Co price on the open market was a major uncertainty. Manganese might be recovered with some possible effects on the world manganese market. Deep-sea mining would have little or no effect on the world copper market. During the next 40 years the situation changed somewhat. At the time of writing, several additional metals in nodules are of economic interest, including Ti, V, Li and the Rare Earth Elements. The main uses of these are in the newly emerged, and emerging, technology areas such as mobile telephones, wind farms, batteries for electric cars, and solar panels, and it is not inconceivable that yet more of the elements in nodules will become of economic value if developing technologies require them.

Postscript: Project Azorian, the Glomar Explorer Affair

Although its contribution to the advancement of deep-sea minerals developments is arguable, the Glomar Explorer affair is worthy of mention if only in that it offers a cautionary tale on how legitimate enterprises can be hijacked for other purposes. In the early 1970s, the ship Glomar Explorer was publicized as a state-of-the-art deep-sea mining vessel designed and owned by the Howard Hughes organization for mining manganese nodules. However, it was actually built to recover a Russian submarine that had sunk in the Pacific some 1600 miles NW of Hawaii.

The CIA had approached Howard Hughes for this task, reportedly because of his earlier dealings with the Agency, and his well-known obsession with secrecy. According to the obituary of the CIA officer tasked with finding the sunken submarine, Christopher Fitzgerald [65], the ship commenced the recovery operation in July 1974, 6 years after the submarine was lost. A retrieval claw was lowered to the sea floor attached to multiple lengths of piping, and part of the submarine was recovered. In the event, the true purpose of the Glomar Explorer was revealed in 1975, after it had completed its task.

Reportedly, associated with the shipboard operation was research funded by the CIA. Several studies carried out in the 1970s received backing, through intermediaries, from Project Azorian (Anon 2023). It could be argued that Project Azorian served to bolster the perceived benefit of deep-sea minerals development in the public eye, and when the true purpose of the operation was revealed left the deep-sea minerals industry somewhat exposed. Project Azorian may even have been detrimental to its development.

References

1. Murray J, Renard AF (1891) Deep-sea deposits. Rep Sci Results Exolor. Voyage HMS Challenger. HMSO, London 525pp
2. Skornyakova NS, Andrushchenko PF, Fomina LS (1962) The chemical composition of iron-manganese nodules in the Pacific Ocean. Okeanologiya 2:264–277. Translated in Deep-Sea Research 11:93–104
3. Arrhenius G, Mero J, Korkisch J (1964) Origin of oceanic manganese minerals. Science 1443:170–173
4. Cronan DS (1967) The geochemistry of some manganese nodules and associated pelagic deposits. Ph.D. Thesis. University of London, Applied Geochemistry Research Group, Imperial College
5. Verlaan P, Cronan DS (2022) Origin and variability of resource-grade marine ferromanganese nodules and crusts in the Pacific Ocean: a review of biogeochemical and physical controls. Geochemistry 82:125721
6. Raitt H (1956) Exploring the deep Pacific. Norton, 272pp
7. Menard H, Shipek C (1958) Surface concentrations of manganese nodules. Nature 182:1156–1158
8. Mero JL (1959) The mining and processing of deep-sea manganese nodules. Report. Institute of Marine Resources, University of California, 96pp

9. Mero JL (1962) Ocean floor manganese nodules. Econ Geol 57:747–767
10. Mero J L (1965) The mineral resources of the sea. Elsevier, Amsterdam, 312pp
11. Zenkevitch N, Skornyakova NS (1961) Iron and manganese on the ocean bottom. Natura (USSR) 3:47–50
12. Cronan DS (1972) Regional geochemistry of ferromanganese nodules in the world ocean. In: Horn DR (ed) Ferromanganese deposits on the ocean floor. National Science Foundation, Washington DC, pp 19–30
13. Andrews JE, Friedrich G et al (1984) The Hawaii-Tahiti transect: the oceanographic environment of manganese nodule deposits in the Central Pacific. Mar Geol 54:109–130
14. Skornyakova NS (1979) Zonal regularities in occurrence, morphology and chemistry of manganese nodules in the Pacific Ocean. In: Bischoff JL, Piper DZ (eds) Marine geology and oceanography of the Pacific Manganese Nodule Province. Plenum, New York, pp 699–727
15. Raab W (1972) Physical and chemical features of Pacific deep-sea manganese nodules and their implications to the genesis of the nodules. In: Horn DR (ed) Ferromanganese deposits on the ocean floor. National Science Foundation, Washington DC, pp 31–50
16. Horn DL, Horn BM, Delach MN (1972) Distribution of ferromanganese deposits in the world ocean. In: Horn DR (ed) Ferromanganese deposits on the ocean floor. National Science Foundation, Washington DC, pp 9–18
17. Mero JL (1978) Ocean mining-an historical perspective. Mar Min 1:243–255
18. Cronan DS (1980) Underwater minerals. Academic Press, 362pp
19. Lipton I, Nimmo, M, Parianos J (2016) TOML Clarion-Clipperton Zone project, Pacific Ocean. NI43-101 report. AMC Consultants, Brisbane. www.sedar.com
20. Parianos J (2017) History of development of the Clarion- Clipperton Zone https://www.lulu.com/shop/john-parianos/history-of-development-of-the-clarion-clipperton-zone/paperback/product-1vj9nm22.html?page=1&pageSize=4
21. Wright R (1976) Ocean mining: an economic evaluation. Ocean mining administration professional staff study 18pp+2 appendices. US Dept of the Interior 1976
22. Earney FCF (1990) Marine mineral resources. Routledge, London, 387pp
23. Usui A, Moritani T (1992) Manganese nodule deposits in the Central Pacific basin: distribution, geochemistry, mineralogy and genesis. In: Keating B, Bolton B (eds) Geology and offshore mineral resources of the Central Pacific Basin, Circum-Pacific Council for energy and mineral resources. Earth science series, vol 14. Springer, New York
24. Glasby GP, Lawrence P (1974) New Zealand Oceanographic Institute Chart Series, pp 33–38
25. Glasby GP (1976) Manganese nodules in the South Pacific: a review. N Z J Geol Geophys 19:707–736
26. Thijssen T, Glasby GP et al (1981) Reconnaissance survey of manganese nodules from the northern sector of the Peru basin. Mar Min 2:385–429
27. Feller R (1987) Manganknollenbergbau-Realitat oder Vision. Beitrage zur Meerestechnik (Clausthal-Zellerfeld) 11:177–198
28. Thijssen T, Glasby GP et al (1985) Manganese nodules in the Central Peru basin. Chem Erde 44:20–27
29. von Stackelberg U (2000) Manganese nodules of the Peru Basin. In: Cronan DS (ed) Handbook of marine mineral deposits. CRC Press, pp 197–238
30. Skornyakova NS, Andrushchenko PF (1970) Iron manganese nodules in the Pacific Ocean In Bezrukov PL (ed) Sedimentation in the Pacific Ocean 2 Nauka (in Russian) translated in Int Geol Rev 16, pp 863–919 (1974)
31. Glasby GP, Exon NF, Meylan MA (1986) Manganese nodules in the S W Pacific. In: Cronan DS (ed) Sedimentation and mineral deposits in the southwestern Pacific Ocean. Academic Press, pp 237–262
32. Kingan SG (2001) Paradise or comic opera. South Pacific Applied Geoscience Commission, Suva, 260pp
33. Glasby GP, Backer H et al (1974) Extensive manganese nodule province discovered in the Southwest Pacific near New Zealand. Meerestechnik 5:145–147

34. Exon NF (1983) Manganese nodule deposits in the Central Pacific Ocean and their variation with latitude. Mar Min 4:79–108
35. Glasby GP, Katz HR (eds) (1976) CCOP/SOPAC tech bull 2. Suva, 165pp
36. Committee for Coordination of Joint Prospecting for Mineral Resources in South Pacific Offshore areas (CCOP/SOPAC). Proceedings of the Fourth Session. Honiara, Solomon Islands, 8–16th September 1975
37. Usui A (1983) Regional variation in manganese nodule facies on the Wake-Tahiti Transect: morphological, mineralogical and chemical study. Mar Geol 54:27–51
38. Cronan DS, Hodkinson R (1994) Element supply to surface manganese nodules along the Aitutaki-Jarvis Transect, South Pacific. J Geol Soc Lond 151:391–401
39. Cronan DS (1977) Deep-sea nodules: distribution and geochemistry. In: Glasby GP (ed) Marine manganese deposits. Elsevier, pp 11–44
40. Cronan DS (1975) Manganese nodules and other ferromanganese oxide deposits from the Atlantic Ocean. J Geophys Res 80:3831–3837
41. Bezrukov PL, Andrushchenko PF (1974) The geochemistry of ferromanganese nodules from the Indian Ocean. Int Geol Rev 16:1044–1061
42. Cronan DS, Tooms JS (1967) Geochemistry of manganese nodules from the N W Indian Ocean. Deep Sea Res 14:239–249
43. Cronan DS (1977) RRS shackleton in the indian and south atlantic oceans. J Geol Soc Lond 1(34):77–80
44. Colley NM, Cronan DS, Moorby SA (1979) Some geochemical and mineralogical studies on newly collected ferromanganese oxide deposits from the N W Indian Ocean. Colloques Internationaux du C.N.R.S. 289-La Genese des Nodules de Manganese, pp 13–21
45. Shama R (2010) First nodule to first mine site: development of deep-sea mineral resources from the Indian Ocean. Curr Sci 99:750–759
46. Siddique HN, Das Gupta DR et al (1978) Manganese-iron nodules from the Indian Ocean. Ind J Mar Sci 7:239–253
47. Frazer J, Wilson L (1979) Manganese nodule deposits in the Indian Ocean. S I O Oceanogr Ref Ser 79–18, 71pp
48. Moorby SA (1978) The geochemistry and mineralogy of some ferromanganese oxides and associated deposits from the Indian and Atlantic Oceans. Ph.D. Thesis. University of London, Applied Geochemistry Research Group, Imperial College
49. Cronan DS, Moorby SA (1982) Manganese nodules and other ferromanganese oxide deposits from the Indian Ocean. J Geol Soc Lond 138:527–539
50. Frazer J (1977) Manganese nodule reserves: an updated estimate. Mar Min 1:103–123
51. Frazer (1982) Assessment of manganese nodule resources. In: Seabed minerals 1. Graham and Trotman
52. Archer AA (1976) Prospect for the exploitation of manganese nodules: the main technical, economic and legal problems. In: Glasby GP, Katz HR (eds) CCOP/SOPAC tech bull 2, Suva, pp 21–38
53. Archer A A (1977) Reserves and potential resources of Ni and Cu in manganese nodules. U N Group of experts report
54. Archer AA (1987) Sources of confusion: what are marine mineral resources. In: Teleki P et al (eds) Marine minerals: advances in research and resource assessment, NATO ASI series C, vol 194. D Reidel, pp 421–432
55. Blondel F, Lasky SG (1956) Mineral reserves and mineral resources. Econ Geol 51:686–697
56. Bastien-Thiry H, Lenoble JP, Rogel P (1977) Manganese nodule resources in the North Pacific. Energy Mining J 178:86
57. Holser AF (1976) Manganese nodule resources and mine site availability. U S Ocean Mining Administration Professional Staff Study 12pp
58. Kildow JT, Bever MB et al (1976) Assessment of economic and regulatory conditions affecting ocean mineral resources development. MIT Report. US Department of the Interior (unpublished)

59. Glasby GP (1983) The three million tons per year nodule mine site:an optimistic assumption. Mar Min 4:73–77

60. Nyhart J D (1978) A cost model of deep ocean mining and associated regulatory issues (MITSG 78-4). In: Methodologies for assessing the impact of deep sea-bed minerals on the world economy. United Nations, 1986

61. United Nations (1986) Methodologies for assessing the impact of deep sea-bed minerals on the world economy. 153pp

62. Levy JP, Odunton N (1984) Natural Resources Forum 8, No 2

63. Nyhart JD et al (1983) A pioneer deep ocean mining venture (MITSG 83-14). In: United Nations 1986, Methodologies for assessing the impact of deep sea-bed minerals on the world economy. United Nations, 1986

64. Flipse J (1982) An economic analysis of a pioneer deep ocean mining venture (R/COE-3 TAMU-SG-82-201). Texas A&M University, College Station

65. Jackson H (2009, November 9) Obituary of Christopher Fitzgerald. CIA officer tasked with finding a sunken soviet submarine. Guardian Newspaper (UK)

Cobalt-Rich Crusts: Recognition and Preliminary Evaluations

Introduction

Cobalt-rich crusts are ferromanganese oxide crusts on hard rock substrates that contain above average Co, up to around 2%, compared with marine ferromanganese oxides as a whole. They became of economic interest in the early 1980s. Prior to the 1980s, nodules and crusts were not always distinguished from each other and some of the nodule analyses recorded in the early literature were probably analyses of crust fragments. In those days most sampling was done by dredge and this could break up nodules as well as break crusts off from their hard rock substrates, mixing them up. Mero [1] never specifically discussed crusts in his book, although he was aware of the difference between crusts and nodules. He commented that of five ferromanganese oxide samples available to him from the South Atlantic, four were crusts or manganese dioxide-impregnated pumice and only one was a nodule. Nevertheless, routine distinction between crusts and nodules was not made for about another 15 years. It was a result of the work of A. Aplin of Imperial College, P. Halbach of Clausthal University, and J Hein and F T Manheim of the USGS in the early 1980s that ferromanganese oxide crusts became generally regarded as a separate class of deposits, at least from the resource point of view. The early history of crust studies had been reviewed by Hein et al. [2].

Regional Studies

Some of the earliest studies on Co-rich crust deposits were made on the Blake Plateau where they are associated with nodules (see Chapter "Activities on Manganese Nodules During The Post-war Boom"), and in the Hawaiian Archipelago [3]. The latter were described by Craig et al. [4]. According to these authors, the

Hawaiian Archipelago (Fig. 1) consists of a linear chain of volcanic features on the Hawaiian Ridge. The sea floor in the Archipelago descends in seven terraces with steep and rugged intervening slopes. Ferromanganese oxide pavements and crusts up to 2–4 cm in thickness were observed on many of the terraces. Manganese was found to range between 15% and 25% in the deposits, whereas Fe averaged about 15–16%. Titanium was generally greater than 0.6%. Copper, Ni, and Co values were found to have a wide scatter with Ni averaging 0.3–0.4% and Co ranging up to more than 1%. Copper was low. These studies on Hawaiian crusts soon led to a consideration of their economic potential [5].

Some of the earliest studies on crusts from elsewhere in the Pacific Ocean resulted from a Hawaii Institute of Geophysics (HIG) cruise to the Line Islands Ridge in 1979 to study basement rocks. Most rock samples collected were heavily encrusted with ferromanganese oxides and these provided material for geochemical studies. These were carried out at Imperial College London in 1981 and 1982 and reported on by Aplin [6] who initially elucidated many of the geochemical features of crusts that are well known today.(see [7] for a review).

In 1981, the German vessel R V Sonne further sampled the ferromanganese oxide crusts of the Line Islands Ridge and also sampled in the Mid-Pacific Mountains during the Midpac 81 cruise (Fig. 2) [8]. According to Glasby [9], this led to the United States Geological Survey taking an interest in these deposits as a possible mineral resource, reporting that it had been suggested by Halbach and Manheim [10] that a single seamount could yield enough ore for a commercial mining operation of up to 4 million tonnes a year. The then Director of the USGS, Dallas L Peck, was reported [11] as saying that crust deposits could warrant commercial

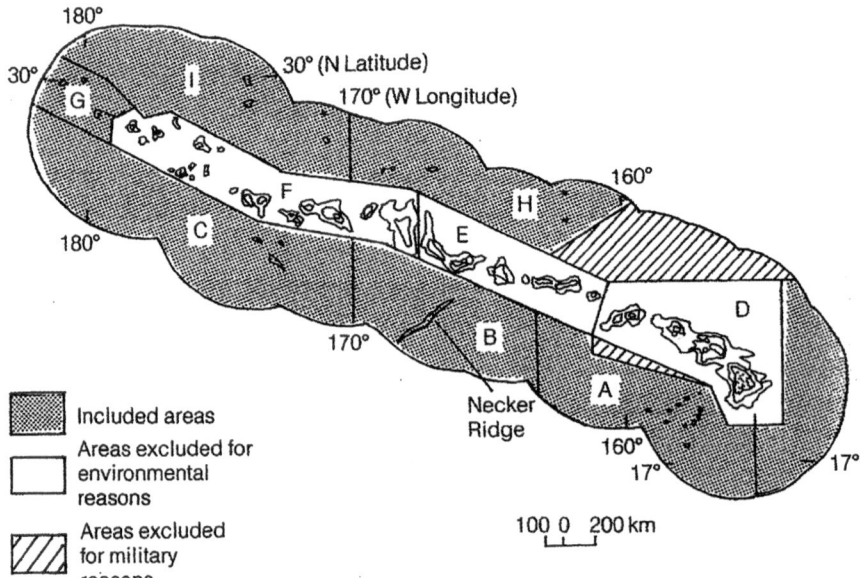

Fig. 1 Areas of Co-rich crust investigations in the Hawaiian Archipelago in the early1980s

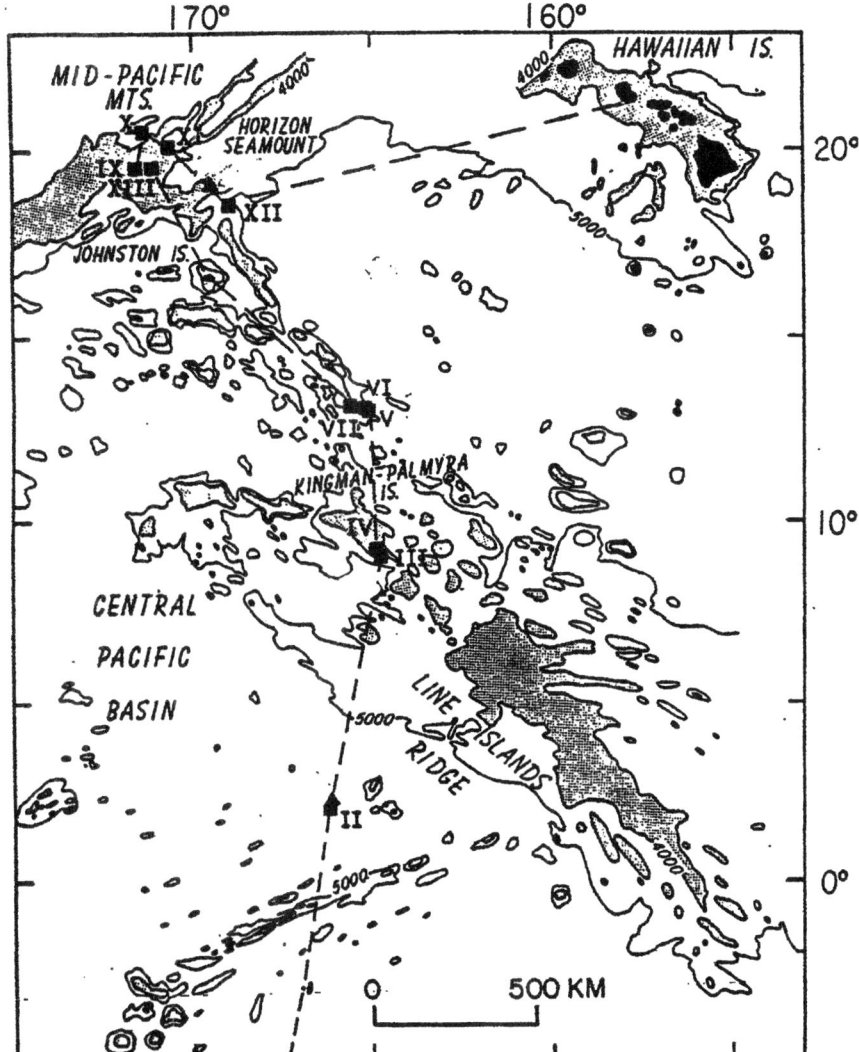

Fig. 2 Location of the MIDPAC 81 cobalt-rich crust expedition showing sample locations II to XIII. (Courtesy of P Halbach)

operations in the future as technology develops. He also suggested that similar cobalt-rich deposits may be available at shallow depth near the US Pacific Trust Territories. There can be little doubt that such sentiments did much to further the USGS Co-rich ferromanganese crust exploration activities of the early and mid-1980s. Glasby [9] considered that crust exploration would have been important to the US at that time as it had no domestic source of Co and that this metal was listed as one of the five most important strategic metals to the USA. Thus, the Midpac 81 cruise of Sonne was very timely in stimulating interest in a possible new marine mineral resource.

Fig. 3 Sample of cobalt-rich ferromanganese oxide crust. (Courtesy of JR Hein)

Crust Characteristics

An example of a cobalt-rich crust is shown in (Fig 3). The growth rates of the crusts were found to be consistent with the rates of slowly growing nodules, i.e. 1–3 mm per million years. Crust thickness was found to increase with substrate age. Because of this, substrates older than 20–30 million years were considered the most likely to be coated with the thickest and therefore the most economically valuable deposits. Two generations of crusts were reported to occur on some of the oldest substrates. The highest concentrations of Co in crusts were found to occur in deposits from less than 2500 m, mainly on seamounts. Island slopes were found to be not so favorable for crust growth because of mass wastage, i.e. the slumping of rock and sediment down the submerged slopes of the islands either burying the crusts or breaking them off. Accordingly, the main areas in which crust exploration was concentrated were the so-called "permissive areas"on plateaux and seamount slopes between about 800-2500 m.

Economic Potential

According to Earney [12], some authorities in the 1980s considered that Co-rich crusts had some economic advantages over nodules and might be mined before them. Several factors led to this appraisal. First, they were found to occur in shallower water areas than manganese nodules, often 2000 m, or more shallower, and in places they occurred within only a few hundred meters of the sea surface. Second, their coverage of the sea floor in some areas was denser than that of nodules in prime nodule areas. Third, Co was viewed as a metal of high strategic value, and which was low in most Ni and Cu-rich nodules. Fourth, by their very occurrence on seamounts, they commonly fell within the jurisdiction of island states, because the seamounts themselves were often just the submerged parts of island chains. Set against these advantages, however, was that because crusts are mostly firmly attached to the sea floor, they would be much more difficult to recover than nodules.

It was difficult to make reliable resource estimates of crusts in the early days of interest in them because of the limited data available. However, assuming a thickness of 2 cm and a bulk wet density of 2, Halbach et al. [8] estimated a crust coverage of 40 kg per sq. m on some seamount slopes in the Pacific. On this basis and the compositional data available on the crusts, these authors considered metal quantities in them to be overall in the same order of magnitude as in nodules from the CCZ. Exploration for and research on Co-rich crusts continued into the 1990s and this will be discussed in Part II of this book.

References

1. Mero JL (1965) The mineral resources of the sea. Elsevier, Amsterdam, 312pp
2. Hein JR, Schulz MS, Gein Lisa M (1992) Central Pacific cobalt-rich ferromanganese crusts: historical perspective and regional variability. In: Keating BH, Bolton BR (eds) Geology and offshore mineral resources of the central Pacific basin, Circum Pacific Council for energy and mineral resources earth science sections, series, vol 14. Springer, New York, pp 261–283
3. Moore JG (1965) Petrology of deep-sea basalt near Hawaii. Am J Sci 263:40–45
4. Craig JD, Andrews JE, Meylan MA (1982) Ferromanganese deposits in the Hawaiian Archipelago. Mar Geol 45:127–157
5. Helsey CE, Keating B, De Carlo E et al (1985) Resource assessment of cobalt-rich ferromanganese crusts within the Hawaiian Exclusive Economic Zone. Final report 14-12-001-30177. US Minerals Management Service
6. Aplin A (1983) The geochemistry and environment of deposition of some ferromanganese oxide deposits from the south equatorial Pacific. Ph.D. Thesis. University of London, Applied Geochemistry Research Group, Imperial College
7. Verlaan PA, Cronan DS (2022) Origin and variability of resource grade marine ferromanganese nodules and crusts in the Pacific Ocean. Geochemistry 82:125741
8. Halbach P, Manheim FT, Otter P (1982) Co-rich ferromanganese deposits in the marginal seamount regions of the Central Pacific Basin-results of the Midpac 81. Erzmetall 35:447–453
9. Glasby GP (1986) Marine minerals in the Pacific. Oceanogr Mar Biol Ann Rev 24:11–64
10. Halbach P, Manheim FT (1984) Potential of cobalt and other metals in ferromanganese crusts on seamounts of the Central Pacific Basin. Mar Min 4:319–336
11. Ocean Industry, 17, October 1982, pp 138–140
12. Earney FC (1990) Marine mineral resources. Routledge, London, 385pp

Hydrothermal Deposits: Discovery and Preliminary Economic Evaluation

Introduction

Mero [1] mentioned the possibility that deep-sea red clays might be mined some-time in the future but made no mention of hydrothermal deposits, even though he was probably aware of their existence, working, as he was, in close proximity to early marine hydrothermalists such as Kurt Bostrom and Enrico Bonatti at the Scripps Institution of Oceanography in the 1960s. Submarine hydrothermal depos-its were initially reported in the Red Sea by Miller et al. [2] and in the Pacific by Bostrom and Peterson [3] and Bonatti and Joensuu [4], and in shallow water off Santorini in the Aegean Sea where iron rich hydrothermal deposits were reported by Butuzova [5].

Actually, hydrothermal mineral deposits had been suspected to occur on the sea floor before they were first found. Sir Kingsley Dunham in his 1968 Presidential Address to the Geological Society of London [6] said "I myself have been willing to admit the possibility of local enrichment of sea water by metals in a restricted environment by ingress of hot hydrothermal water" [7]. Following the initial discov-eries, submarine hydrothermal deposits were found to be quite widespread and more variable in composition than ferromanganese oxide nodules and crusts. Indeed, it soon became apparent that ferromanganese oxides themselves could be hydrothermal in origin (see [8] for a review of early findings). Hydrothermal miner-als that were found during early phases of investigation of the deposits included oxides, sulphides, and silicates of Fe, Mn, Cu, Zn, and sometimes Pb, together with compounds of more valuable metals such as Au and Ag.

D. S. Cronan, *Deep-Sea Minerals Developments in the 20th Century*,
https://doi.org/10.1007/978-3-031-52342-7_4

Red Sea Deposits

Hydrothermal deposits of potential economic value were first found in a number of deeps in the median valley of the Red Sea. The earliest indications that hydrothermal activity was occurring on the floor of the Red Sea were high temperatures and salinities in the bottom waters of the median valley recorded by the Albatross in 1948 during the Swedish Deep-Sea Expedition [9].[1] These indications were confirmed some years later when further measurements were made by Woods Hole ships en-route to the Indian Ocean to participate in the International Indian Ocean Expedition, and the area was subjected to a detailed survey by the Woods Hole Oceanographic Institution during several cruises between 1963 and 1967. At this time metal-rich sediments and brines were sampled [2]. The work was published in a volume edited by Degens and Ross [10] which is still a standard text on the subject. Subsequent work was carried out by other groups and these investigations resulted in the finding of about a dozen brine pools and a number of areas floored by metalliferous sediments [11] (Fig. 1). Most of the Red Sea hydrothermal mineralization occurs in its central and northern parts where the median valley is well developed. According to Bignell [12, 13], the major deeps in this area occur where transform faults intersect the median valley. He suggested that the discharge of metal-rich brines from which many of the hydrothermal sediments are deposited is controlled by this transform faulting.

One of the most characteristic features of Red Sea hydrothermal sediments first noticed in the Atlantis II Deep, and which became apparent quite early on in the investigation of them was the considerable diversity that they often exhibit both vertically and laterally [14]. Investigations on metalliferous sediments from other deeps [11] resulted in their overall classification into five major groups, oxides, sulphides, sulphates, silicates and carbonates, and several sub-groups. The sulphides, oxides and silicates were the sediments with the greatest metal enrichments, and the sulphides were subdivided into a further two sub-groups, mixed sulphides of Fe-Cu-Zn, and pyrite (iron disulphide). Of these, the Fe-Cu-Zn-rich sulphides were the ones considered to be of the most economic value. These hydrothermal-sedimentary precipitates were found to have a high brine-fluid content and a muddy consistency; any consolidation had not yet taken place.

Of all the metalliferous sediments discovered in the Red Sea, only those in the Atlantis II Deep were thought likely to be of economic importance at the time. The Atlantis II Deep is situated in the median valley of the Red Sea at about 21.5 N,38 E in an area of irregular fractured topography. It is elongated and covers an area of about 60 sq. km. Hydrographic investigations showed the presence of two brines

---

[1] In the early 1970s the geochemist on the Swedish Deep-Sea Expedition, Gustav Arrhenius, told the writer that had he been on board the Albatross during its transit of the Red Sea, he would have asked for a piston core to be taken at the site of the water anomalies, and the Red Sea hydrothermal deposits would most likely have been discovered in 1948. As it was, Arrhenius left the ship before the Red Sea transit to go back to Sweden and did not rejoin the ship until after it was complete.

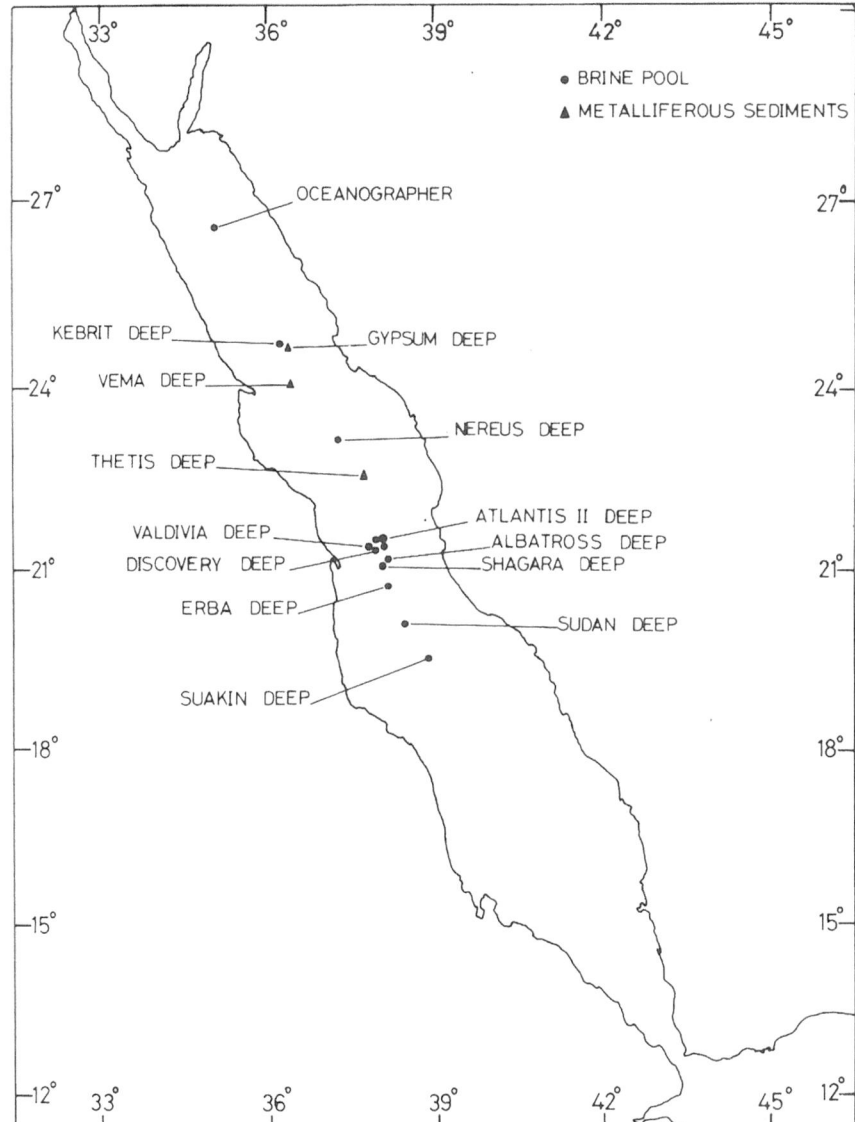

Fig. 1 Location of brine pools and metalliferous sediments in the Red Sea in 1975

within it, a lower one that had a salinity of 25.7% and a temperature of 56.5 °C in 1966 which rose to 60 °C by 1972, and an upper one with a salinity of 13.5% and a temperature of 44.3 °C in 1966 which had risen to 50 °C in 1972. The brines were enriched relative to seawater in Fe, Mn, Zn and Cu [15]. Consideration of the deposits as an ore body was based on contents of 2–6% Zn 0.5–1% Cu and 50–100 g/tonne of Ag and 0.5 g/tonne of Au, with an anticipated production of 2,600,000 tonnes per year of mined sediment yielding 90,000 tonnes of 32% Zn concentrate, and corresponding amounts of Cu, Ag and Au [16].

An early estimate of the economic worth of the Red Sea hydrothermal deposits was made by Hackett and Bischoff [17]. For this purpose, the Atlantis II Deep was divided into three areas representing partially separate basins. The value of the deposits was estimated to be $2.33 billion at 1972 prices. Backer [18] estimated that the Deep contained about 2.5 million tonnes of available Zn, with 0.6 mt of Cu and 0.009 mt of Ag.

According to Mero [19], in 1967 the Sudanese Govt laid claim to the deposits and granted a permit to explore them further to Sudanese Minerals Ltd., a subsidiary of the International Geomarine Corp of Los Angeles. The following year, Preussag AG became involved in the venture. Later, a coup in Sudan introduced a new government which changed the conditions of the exploration agreement to one that gave Sudan 90% of net profits. As a result of this, the deposits ceased to be of economic interest to Sudanese Minerals Ltd. and they withdrew from the project. In 1975, with the aid of Saudi Arabia's financial resources, the Red Sea Commission was formed to finance the development of the Red Sea deposits. Much of the subsequent work on them was undertaken under the aegis of that Commission. Preussag AG was appointed as its contractor. They determined that the mining of the deposits was technically possible and the economics of the operation appeared competitive with onshore mining operations and could be performed in an environmentally acceptable manner [16]. Tests determined that ore processing and disposal of waste at sea would be necessary to meet cost objectives. Ultrafine particle processing aboard ship proved to be possible. Such flotation provided a bulk sulphide concentrate. However, in spite of successful tests the mining of the Red Sea metalliferous sediments did not go ahead in the twentieth century but is being actively reconsidered at the time of writing (see Epilogue).

Polymetallic Sulphide (PMS) and Related Deposits

The discoveries of hydrothermal metalliferous sulfide muds in the Red Sea in the mid-1960s prompted a search of other submarine volcanic areas where similar deposits might occur. In 1975 the RV Dmitri Mendeleev dredged a lithified pelagic turbidite in the Hess Deep in the eastern Pacific Ocean at a depth of between 5100 m and 5280 m. It contained hydrothermal talc, smectite and pyrrhotite which was enriched in Cu (0.38%), and Zn (0.02%) [20].

Lonsdale et al. [21] described material recovered in November 1977 by the US Navy's submersible Seacliff, collected during four dives on the sea floor spreading axis in the Guaymas Basin in the Gulf of California. The submersible observed extensive terraces of talc, samples of which contained hydrothermal smectite and pyrrhotite. The talc was iron-rich. Isotope studies indicated a precipitation temperature of about 200 C. Later, sediment-covered massive sulphides were located in the Gulf of California [22].

In the late 1970s, sulphides rich in Cu, Zn and Fe were found on the EPR at 21 N [23]. They were located in a depression about 200 m west of the mid-ocean ridge extrusion zone less than 1 km from the spreading axis. Similar deposits were subsequently found to occur along several km of ridge crest in the same general area. Speiss et al. [24] reported the finding of hot springs between March and May 1979 on the EPR near 21 N using the Woods Hole Oceanographic Institution Alvin submersible. Vents with water as hot as 380 C were found as well as cooler springs surrounded by dense biological communities.

Submarine hydrothermal activity was also found in 1979 off the Galapagos Islands [25] Two types of hydrothermal activity were observed, hot springs at five locations in the Galapagos Rift and hydrothermal sediment mounds 20 km to the south. Submersible operations with the WHOIs Alvin were carried out in both areas. The hydrothermal vents were clearly discernable by the abundant remains of organisms around them. Also, streams of milky water were seen discharging from the vents. The water temperature was 10–11 °C at the mouth of one of the vents, while 40 cm above it had fallen to 4–6 °C on mixing with seawater. There was no oxygen in the fluids but most contained hydrogen sulphide. Several elements were determined in the vent discharges including Fe, Mn, Ni, Cu and Cd. Iron was high in the waters from one of the vents, but not the others, possibly suggesting its subsurface precipitation. Sulphide precipitation on the sea floor was observed in the form of sulphide pinnacle-like chimneys. The hydrothermal mounds to the south of the vent area were found to contain sharply differentiated iron oxides, manganese oxides and iron silicates, and were saturated with warm water.

One of the largest of the early PMS discoveries was also in the Galapagos Rift, at 86 W, where Alex Malahoff of the US National Oceanic and Atmospheric Administration (NOAA) reported finding an inactive PMS deposit over 1 km long and 100–200 m wide at a depth of 2600 m [26]. Thickness was not confirmed by drilling but it was estimated to be up to 12 m. On this basis, Malahoff considered that there was a total of around 25 million tonnes of sulphide minerals present in it, valued at something near 2 billion dollars at 1982 prices. Minerals found in this deposit included pyrite, sphalerite, chalcopyrite, wurtzite and marcasite and the grades were Cu, 0.3–11% and Zn, 0.5–50%.

Hydrothermal PMS deposits were found in the early 1980s on the Juan de Fuca Ridge near the coast of the USA and were described by Koski et al. [27]. In 1981 geologists from the USGS and University of Washington dredged PMS at about 2200 m about 500 km off the Oregon coast at the southern end of the Juan de Fuca Ridge. The USGS reported finding a zinc-iron sulphide [28]. Like the Galapagos deposits, this deposit was inactive but showed evidence of being historically recently active by the presence of remains of marine fauna often associated with submarine hydrothermal activity. A nearly 10 kg sample containing 30–54% Zn sulphide was recovered. Coincident with the finding of these deposits, NOAA was reported to be intending to launch a five-year programme to determine the extent of hydrothermal mineralisation off the US west coast. As part of this, the NOAA ship RV Surveyor carried out a research cruise in March 1982 to do a high-resolution bathymetric survey and collect samples. One of the purposes of the programme was to determine the potential for mining the deposits prior to a possible lease sale. Of course, it was far too early to say if any of the deposits were of economic value and would ever be mined.

Throughout the 1960s and 1970s, the main effort in deep-sea hydrothermal mineral exploration had been in the Red Sea and in the eastern Pacific. In the western Pacific, there are also spreading ridges in back-arc basins, which also have the potential to host hydrothermal fluids and associated mineralization. These were thought to have the potential to host PMS deposits but none were found until well into the 1980s (see Part II). However, indications of their possible presence were seen in occurrences of ferruginous sediments described from the Lau Basin by Griffin et al. [29] and Bertine [30], and in barite of supposed hydrothermal origin described by Bertine and Keen [31] from a fracture offsetting a spreading centre there. The discovery of hydrothermal manganese oxide on the Tonga -Kermadec Ridge in an island arc setting in 1981[2] [32] by the NZOI research vessel, RV Tangaroa, indicated that island arcs might also have the potential to host PMS deposits, as manganese oxides had been found to be associated with hydrothermal sulphides in the Red Sea. The veracity of this supposition was borne out by the finding of several such deposits on the TKR and other island arcs in the subsequent decades.

The early discoveries of hydrothermal mineral deposits in the western Pacific outlined above, led CCOP/SOPAC in 1981 to commence a study of all likely locations for their occurrence in its area. Based largely on a study of the geochemistry of surface sediments in the southwestern Pacific, Cronan [33] suggested the possibility of hydrothermal deposits in two geological settings (a) near proposed extensional or transform plate boundaries and (b) associated with single volcanoes in zones of convergence. Specific areas thought to have a potential for hydrothermal deposits included the Bismarck Sea, the Woodlark Basin, the North Fiji Basin, the Lau Basin, the submarine volcanoes off Vanuatu, and the Tonga-Kermadec Ridge (Fig. 2).There have been hydrothermal discoveries throughout these regions, but the only mining licence for PMS that has been issued there was in the Bismarck Sea.

Hannington et al. [34, 35] have further reviewed hydrothermal PMS deposits.

Metalliferous Sediments

Metalliferous sediments rich in Fe, Mn and several other elements were found in association with submarine volcanic activity from the 1960s onwards. Such deposits were first found near the crest of the EPR by Bostrom and Peterson [3] and on its flanks just above basement by the Deep Sea Drilling Project soon after. They form as a result of the settling of particles from hydrothermal plumes. The deposits occur

---

[2]This deposit was found by a young New Zealand PhD student, Karin Knedler. She was working the midnight to 8 am watch on the Tangaroa and deployed the dredge at a location she selected from the underway PDR records. This was the first time that hydrothermal deposits had been recovered from an island arc.

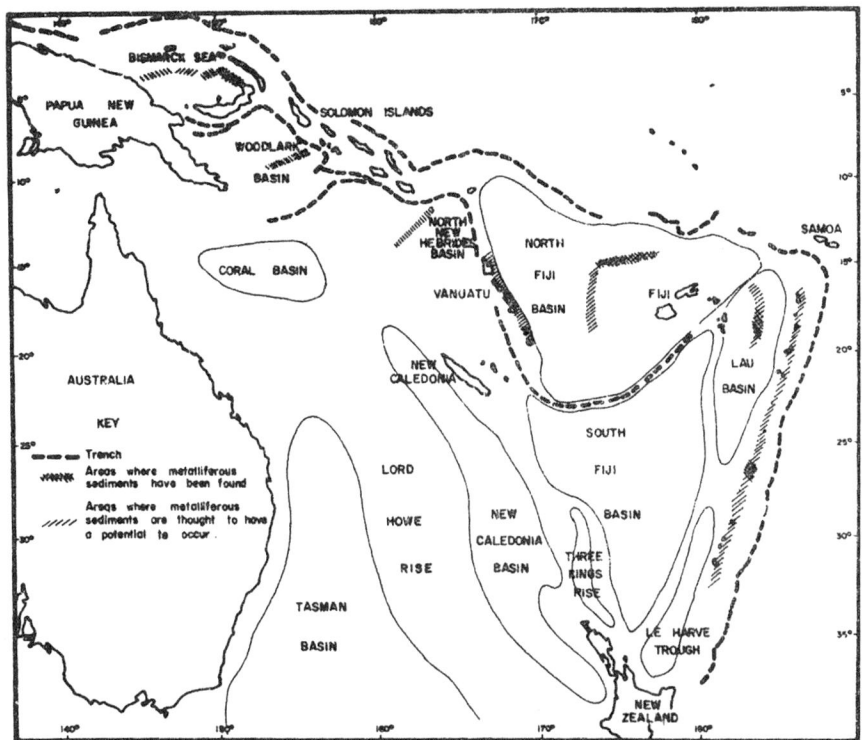

Fig. 2 Known and predicted locations of hydrothermal deposits in the SW Pacific in 1981

intermittently all along the WMORS and also occur in association with submarine volcanism at convergent plate margins. They were also found to be associated with the hydrothermal deposits in the Red Sea where they sometimes form a halo of Fe and Mn enrichment in sediments around the hydrothermal PMS deposits [34]. These deposits were (and are) thought to be of little or no economic value in their own right. However, their exploration value was first recognised in the Red Sea by Bignell et al. [36] and thought to be that their presence on the sea floor pointed to the possible presence of PMS deposits in their vicinity, as occurs in the Red Sea. Subsequently, searching for "hydrothermal haloes" in exploration for hydrothermal minerals on the sea floor became an accepted method of seafloor PMS prospecting (see Chapter "Exploration and Mining Development").

Hydrothermal Ferromanganese Oxide Crusts

Hydrothermal ferromanganese oxide crusts, very rich in Mn, were, as mentioned, found on the WMORS in the 1960s [4], but were not considered to be of any economic value. Like metalliferous sediments, their importance was thought to be that their presence indicated the possible occurrence of higher-grade more economically valuable PMS deposits in their vicinity. Many more submarine hydrothermal fer-romanganese oxide deposits were found in later decades and some found near Pitcairn Island, Britain's last remaining colony in the South Pacific, were cited by the UK Government in justification for an EEZ to be declared around it [37].

References

1. Mero JL (1965) The mineral resources of the sea. Elsevier, 312 pp
2. Miller AR, Densmore CD et al (1966) Hot brines and recent iron deposits in deeps of the Red Sea. Geochim Cosmochim Acta 30:341–359
3. Bostrom K, Peterson MNA (1966) Precipitates from hydrothermal exhalations on the East Pacific Rise. Econ Geol 61:1258–1265
4. Bonatti E, Joensuu O (1966) Deep-sea iron deposits from the South Pacific. Science 157:643–645
5. Butuzova G, Yu (1966) Iron ore sediments of the fumarole field of the Santorini volcano, their composition and origin. Dokl Acad Nauk SSSR 168(6):215–217
6. Dunham KC (1969) Practical geology and the natural environment of man II: seas and oceans. Q J Geol Soc Lond 124:101–129
7. Dunham KC (1964) Neptunist concepts in ore genesis. Econ Geol 59:1–21
8. Cronan DS (1980) Underwater minerals. Academic Press, 362pp
9. Bruneau L, Jerlov N, Koczy F (1953) In: Reports of the Swedish Deep-sea Expedition. Phys Chem 3
10. Degens ET, Ross D (1969) Hot brines and recent heavy metal deposits in the Red Sea. Springer, New York, 600pp
11. Bignell RD, Cronan DS, Tooms JS (1976) Red Sea metalliferous brine precipitates. In: Strong D (ed) Metallogeny and plate tectonics, Geol Assoc Canada Spec Paper, vol 14, pp 148–184
12. Bignell R (1975) The geochemistry of metalliferous brine precipitates and other sediments from the Red Sea. Ph D Thesis. Applied Geochemistry Research Group, Imperial College, University of London
13. Bignell R (1978) Genesis of Red Sea metalliferous sediments. Mar Min 1:212–220
14. Bischoff J (1969) Red Sea geothermal brine deposits: their mineralogy, chemistry and gen-esis. In: Degens ET, Ross D (eds) Hot brines and recent heavy metal deposits in the Red Sea. Springer, New York, pp 368–401
15. Hartman M (1985) Atlantis II deep geothermal brine system. Chemical processes between hydrothermal brines and Red Sea deep water. Mar Geol 64:157–177
16. Amman H (1985) Development of ocean mining in the Red Sea. Mar Min 5:103–116
17. Hackett and Bischoff (1973) New data on the stratigraphy, extent, and geologic history of the Red Sea geothermal deposits. Econ Geol 68:244–256
18. Backer H (1979) In: Offshore Mineral Resources. Documents BRGM No 7. Orleans, pp 319–338
19. Mero JL (1978) Ocean mining-an historical perspective. Mar Min 1:243–256
20. Murdma IO, Rosanova TV (1976) Hess deep bottom sediments. In: Geological-geophysical researches in the southeastern part of the Pacific. Nauka, Moscow, pp 252–260. (in Russian)
21. Lonsdale P, Bischoff JL et al (1980) A high-temperature hydrothermal deposit on the seabed at a Gulf of California spreading Centre. Earth Planet Sci Lett 49:8–20
22. Von Damm KL, Edmond J et al (1985) Chemistry of submarine hydrothermal solutions at Guaymas Basin, Gulf of California. Geochim Cosmochim Acta 49:2221–2237
23. Francheteau J, Needham HD et al (1979) Massive deep-sea sulphide ore deposits discovered on the East Pacific rise. Nature 277:523–528

24. Speiss FN, Johnson DR, Macdonald KC et al (1980) Hydrothermal vents on the mid-ocean ridge: an update. Science 207(4438):1421–1433
25. Corliss J (1979, March 16) Submarine thermal springs on the Galapagos rift. Science Mag
26. Malahoff A (1982) A comparison of the massive polymetallic sulphides of the Galapagos rift with some continental deposits. Mar Tech Soc J 16(930):39–45
27. Koski RA, Normark WR et al (1982) Metal sulfide deposits on the Juan de Fuca ridge. Oceanus 25:42–48
28. Anon (1982) Strateg Miner Manage 1(13)
29. Griffin J, Koide M et al (1972) Sediments of the Lau Basin-rapidly accumulating volcanic deposits. Deep-Sea Res 19:139–148
30. Bertine KK (1974) Origin of Lau Basin rise sediment. Geochim Cosmochim Acta 38:629–640
31. Bertine KJ, Keen J (1975) A hydrothermal barite from the Lau Basin. Science 188:150–152
32. Cronan DS, Glasby GP, Knedler K et al (1982) A submarine hydrothermal manganese deposit from the south-West Pacific Island arc. Nature 298:456–458
33. Cronan DS (1983) Metalliferous sediments in the CCOP/SOPAC region of the southwest Pacific, with particular reference to geochemical exploration for the deposits. CCOP/SOPAC Tech Bull 5. Suva Fiji
34. Hannington MD, Petersen S et al (2004) A global database of seafloor hydrothermal systems. Geol Survey of Canada Open File 4598, 9pp
35. Hannington MD, Jamieson JW et al (2010) The abundance of seafloor massive sulfide deposits. Geology 38(11):1023–1026
36. Bignell RD, Cronan DS, Tooms JS (1976) Metal dispersion in the Red Sea as an aid to marine geochemical exploration. Trans Instn Min Metall B 85:273–278
37. Cronan DS (2015) Deep-sea minerals. Geoscientist 28(8):10–15

Phosphorites

Introduction

Submarine phosphorites were first recovered from the Agulhas Bank off South Africa during the Challenger Expedition [1]. Subsequent discoveries in the twentieth century have been reviewed by, among others, Gulbrandsen [2], Tooms et al. [3], Bezrukov and Baturin [4], and Cronan [5]. A variety of fluorapatite called francolite is the principal phosphate mineral present in them. Their main perceived resource potential lay in their possible use as a fertilizer, although it was thought that they might also have applications in the chemical industry [3] because of their sometimes high contents of elements other than P [5].

Phosphate-based fertilizers are required in the production of basic food crops. The growth and consumption of fertilizer production after the Second World War was greater than at any time before that. According to Burnett and Lee [6], production levels in the mid-1970s were five times as high as in the mid-1950s. While pointing out that there has never been a shortage of phosphate fertilizer from on-land sources, Cronan [7] suggested that marine phosphates might be considered locally as a resource if offshore supplies were available nearby (eg New Zealand), while land supplies were far away and costly, or if phosphate-bearing lands were required for uses other than phosphate recovery (eg SE United States).

Occurrence and Associations

Submarine phosphorites generally occur in water depths down to about 1000 m on continental margins, and thus not all can be regarded as deep-sea minerals. They also sometimes occur on seamounts, where they may form a substrate for Co-rich crust accumulation (see Chapter "Cobalt Rich Crusts: Recognition and Preliminary

D. S. Cronan, *Deep-Sea Minerals Developments in the 20th Century*,
https://doi.org/10.1007/978-3-031-52342-7_5

Evaluations"). They reach their greatest abundance off the western margins of the continents (Fig. 1) generally under tropical or sub-tropical waters. Early modern descriptions of phosphorites off NW Africa were made by Summerhayes [8], off SW Africa by Baturin [9] and Price and Calvert [10], off Peru and Chile by Veeh et al. [11] and Burnett [12, 13] and off Southern California by Emery [14] and Mero [15]. They have also been described from some eastern continental margins (Fig. 1), including the south-eastern United States [16] and New Zealand [18]. Seamount occurrences have been mainly described from the Pacific [19]. Murray and Renard [1] thought that the phosphorite deposits they found off southwest Africa were formed by organic processes related to mass fatalities of marine organisms, and most subsequent theories of phosphorite origin have invoked organic processes.One of the most common oceanographic associations of these various deposits was with areas of upwelling currents (bringing nutrients into surface waters) and resulting high biological productivity. A recent review of phosphorites in relation to upwelling has been given by Summerhayes [20] The southwest African phosphorites were found to be more or less confined to the inner shelf, and consist mainly of phosphatized clots of carbonate mud. Their P content reaches around 6% in these clots but up to 25% in associated brittle concretions, illustrating increasing P content with increasing lithification. The NW African phosphorites were found to be submarine outcrops of the phosphorite on the adjacent landmass. They consist of pelletal, conglomeratic, and massive varieties. However, glauconitic phosphorites also occurred in both regions. Phosphorites off western South America were found to occur in two bands in water depths shallower than 1000 m and contain about 10–20% P. The phosphorites off southern California were found to occur as nodules that range in shape from flat slabs to irregular masses; as off the African margin, some of these are glauconitic. Seamount phosphorite exhibits a variety of forms.

The phosphorite deposits on the Chatham Rise east of New Zealand have been the subject of the greatest interest over the longest period. Cullen et al. [19, 21] described them as loose nodular gravel sometimes more than 70 cm thick containing 9.4% P on average. Their origin was described by von Rad and Kudrass [22] as a multi-stage process.

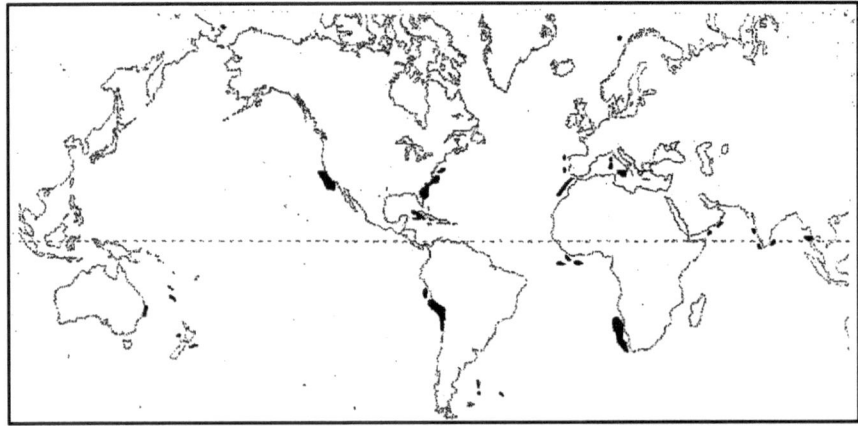

Fig. 1 Worldwide continental margin phosphorite occurrence

Resource Potential

The Pacific Region provided a good example of problems related to coordinating phosphate supply and needs in the 1960s and 1970s. In the 1970s, all the countries of the Asia-Pacific region together produced only about 8% of the world's phosphate supplies but contained 20% of the world's accessible land area and 52% of its population [6]. Japan imported phosphate from Florida while Australia and New Zealand consumed almost the entire output of Nauru and Christmas Island (both subsequently exhausted). Such long supply routes resulted in high transportation costs.

Long supply routes were at least partly why the deposits on the Chatham Rise east of New Zealand (Fig. 2) were suggested as an alternative phosphate source for New Zealand farms. In order to evaluate their potential, a joint New Zealand-West German research program was carried out in 1981 to investigate the possibility of the Chatham Rise deposits being mined at some time in the future [21]. However, Falconer [23] reappraised downward the resource potential of Chatham Rise phosphorite, coinciding with a decline in fertilizer imports into New Zealand after 1985. The company that held the Chatham Rise exploration license allowed it to lapse. However, Falconer [23] considered that should phosphate demands increase, with a concomitant rise in its price, the Chatham Rise deposits might become economic. In fact, the deposits were not mined in the twentieth century, nor were any other marine phosphorites.

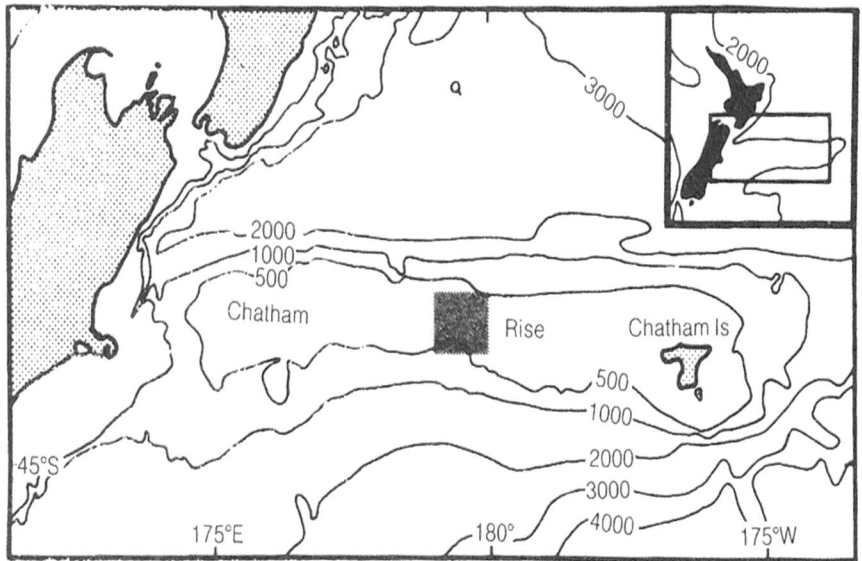

Fig. 2 Area of phosphorite occurrence on the Chatham Rise (references in text)

References

1. Murray J, Renard AF (1891) Deep sea deposits. Report of the scientific results of HMS Challenger 1873–1876. HMSO, London
2. Gulbrandsen J (1966) Chemical composition of phosphorites of the Phosphoria formation. Geochim Cosmochim Acta 30:769–778
3. Tooms JS, Summerhayes CP, Cronan DS (1969) Geochemistry of marine phosphate and manganese deposits. Oceanog Mar Biol Ann Rev 7:49–100
4. Bezrukov P, Baturin GH (1976) Lithology of oceanic phosphorites. In: Lithology of phosphorite bearing deposits. Nauka, Moscow, pp 20–28
5. Cronan DS (1980) Underwater minerals. Academic Press, 362 pp.
6. Burnett WC, Lee AIN (1980) The phosphate supply system of the Pacific Region. Geo J 4(5):423–436
7. Cronan DS (1992) Marine minerals in exclusive economic zones. Chapman and Hall, 209 pp.
8. Summerhayes CP (1970) Phosphate deposits on the Northwest African Continental shelf and slope. Ph D Thesis. University of London, Applied Geochemistry Research Group, Imperial College
9. Baturin GH (2000) Formation and evolution of phosphorite grains and nodules on the Namibian Shelf, from recent to pleistocene. In: Glenn CR, Prevot-Lucas L, Lucas J (eds) Marine authigenesis: from global to microbial. Tulsa: society of economic paleontologists and mineralogists, SEPM special publication, vol 66, pp 185–199
10. Price N, Calvert S (1978) The geochemistry of phosphorites from the Namibian shelf. Chem Geol 23:151–170
11. Veeh HH, Burnett WC, Soutar A (1973) Contemporary phosphorites on the continental margin of Peru. Science 181:844–845
12. Burnett (1974) Uranium-series disequilibrium studies in phosphorite nodules from the west coast of South America. Hawaii Inst Geophys Rept 74–3
13. Burnett (1977) Geochemistry and origin of phosphorite deposits from off Peru and Chile. Geol Soc Am Bull 88:813–823
14. Emery (1960) Geology of the sea floor off California. Wiley, 366 pp.
15. Mero JL (1965) The mineral resources of the sea. Elsevier, 312pp
16. Woolsey Jr. (1976) Neogene stratigraphy of the Georgia coast and inner continental shelf. PhD dissertation, Univeristy of Georgia
17. Riggs SR (1987) Model of tertiary phosphorites on the World's continental margins. In: Teleki P et al (eds) Advances in research and resource assessment, NATO ASI Series. D Riedel, pp 99–118
18. Pasho (1976) Distribution and morphology of Chatham Rise phosphorites. N Z Oceanographic Inst Memoir 77: 28pp
19. Cullen DJ (1986) Submarine phosphatic sediments of the S W Pacific. In: Cronan DS (ed) Sedimentation and mineral deposits in the southwestern Pacific Ocean. Academic Press, pp 183–236
20. Summerhayes CP (2016) Upwelling. In: Harff J, Meschede M, Petersen S, Thiede J (eds) Encyclopedia of marine geosciences. Springer, Dordrecht, pp 900–912
21. Cullen D J, Kudrass H, von Rad U (1981) Preliminary results of the 1981 Sonne investigation of Chatham Rise phosphorite deposits east of New Zealand. In: Proceedings of the international ocean 1981 conference, Dusseldorf, 301/1p-6
22. Von Rad U, Kudrass H (1984) Phosphorite deposits on the Chatham Rise, New Zealand. Geol Jahrb D 65
23. Falconer RKH (1989) Chatham Rise phosphates, a deposit whose time has come (and gone). Mar Min 8:55–67

Exploration and Mining Development

Exploration

Until the 1960s, few attempts were made to explore for deep-sea minerals in a systematic way. The chance nature of the discoveries before that is illustrated by the finding in 1948 of the temperature anomalies indicating the presence of metalliferous brines and sediments in the Red Sea [1].

Visual Methods

Direct visual exploration for deep-sea minerals is only possible from a manned submersible, and the first, time one of these was used, the Nautile, was in the late 1980s in the CCZ. Commonly used in the 1960s and 1970s was photography and later TV. One of the first attempts to use undersea photography in sea floor surveying for manganese nodules was that of Menard and Shipek [2] in their discovery of extensive manganese nodule deposits in the South Pacific. Cameras were designed that could take a sequence of pictures of the sea floor simply by being towed just above the bottom and triggered periodically [3]. Kennecott introduced a stereo camera system in 1972 to determine the size and frequency of seafloor features that were too small to be characterized by PDR. (D Felix pers. comm. 2023). Stereo allowed 3-D resolution on a cm-scale which facilitated mapping the microtopography of the seabed (J Halkyard pers. comm. 2023). Free fall cameras were also designed that fell to the sea floor under their own weight, took a picture just before they hit (Fig. 1), and returned to the surface after discharging a ballast weight. Many of them could take a sample at the same time (see below). Free fall cameras were in widespread use in manganese nodule exploration by the 1980s. Useful as they were, however, cameras did not allow exploration to be carried out in 'real-time'; underwater TV was developed for this.

D. S. Cronan, *Deep-Sea Minerals Developments in the 20th Century*,
https://doi.org/10.1007/978-3-031-52342-7_6

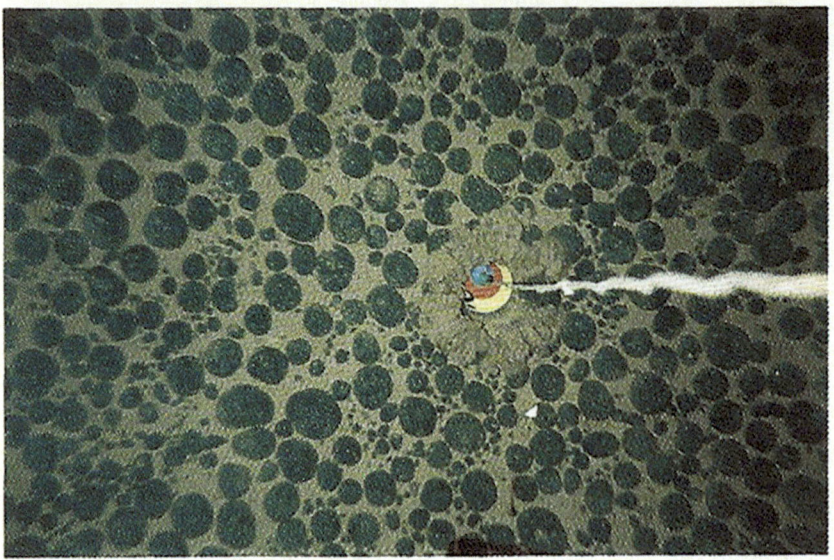

Fig. 1 Free fall camera trigger weight and Pacific seafloor manganese nodules, 1986

Underwater TV had the advantage over still photography in that it allowed the sea floor to be monitored continuously in order to direct exploration activities in real-time and came into use in manganese nodule exploration in the late 1960s. Kaufman and Siapno [4] reported continuous underwater TV operations for up to 5 days in water depths of 4000–5000 meters. However, TV had its limitations such as a restricted field of view due to the low power of the lights on the early cameras and the slow speed at which they had to be towed, leading to the consumption of large amounts of shiptime. Nevertheless, TV was the 'workhorse' in the exploration for manganese nodules in the late 1960s and throughout the 1970s, and provided the equivalent of thousands of still photos. Nevertheless, according to D Felix (pers. Comm. 2023), TV (and bottom photography) were poor methods for determining nodule abundance because many nodules could be partially or completely buried.

Sampling

In order to properly evaluate deep-sea mineral deposits, samples of them had to be obtained. The earliest deep-sea mineral sampling tool to be used was the dredge, which had been in use in sampling the deep-sea floor since the Challenger Expedition in the 1870s. Dredges were routinely used in deep-sea mineral exploration up to the 1970s.[1] Different types of dredges and their limitations have been described by

---

[1] Dredges are still used at the time of writing in order to obtain large bulk samples of nodules or other minerals for processing and other experiments

Cronan [5]. Partly to get over one of their most serious limitations, the sometimes lack of representativity of the material recovered, various types of seafloor grab were designed. However, because of the generally small size of the early grabs, wire-lowered grabs did not find widespread use in the initial phases of deep-sea mineral exploration. It was not until the advent of large hydraulic grabs, in the 1980s, that they came into more common use (see Chapter "Technological Developments 1980–2000"). In the meantime, various varieties of free-fall grabs were developed in the 1970s (Fig. 2). These fell to the sea floor with a ballast weight which was discharged when the grab closed to take its sample, and the grab was returned to the surface by floats. The use of free-fall grabs greatly increased the amount of material that could be obtained from the deep-sea floor in any given time span because several could be deployed at the same time. Kennecott maximized the number of grabs handled per day by mobilizing a helicopter from the vessel that could grapple the FFGs from the water when they returned to the surface (all locatable by radio signal). This meant the vessel could continue on course to drop samplers without having to backtrack (J Halkyard, pers comm. 2023). They allowed samples to be more precisely located than was possible with dredges, and when a camera was attached (Fig. 1) allowed abundance to be calculated as well as providing a sample for chemical analysis. However, according to D Felix (pers. comm 2023) free fall grabs were notoriously inaccurate when used to determine nodule abundance, showing abundances lower than the abundance determined from the camera attachment or by other methods. This might have been because the initial impact of the grab on the sea floor (before it closed) pushed some nodules aside.

If a sample of buried material was required, recourse was made to coring. However, coring achieved little usage in early deep-sea mineral exploration as the deposits sought were almost entirely on the surface of the sea floor. It was only in the Red Sea that the mineral deposits sought were buried and these were sampled initially by piston corer (see [5] for this and other early deep-sea minerals sampling tools). It was not until the advent of box coring in the 1970s that coring was used more routinely in deep-sea mineral sampling. Box corers consisted of a metal box of approximately one-meter square that was pushed slowly into seafloor sediments by a heavy weight and then closed underneath by a metal plate that was swung into position. Its main advantage over other sampling techniques was that it could recover an undisturbed sample of the sea floor (Fig. 3) from which nodule abundance could be accurately estimated.

Remote Methods

Precision depth recording was well established before the advent of deep-sea mineral exploration, but underwent adaption as it developed in the 1960s and 1970s. In addition to sounding the sea floor, PDRs were used in the Red Sea to locate midwater reflectors between different brines or between brine and seawater overlying the Red Sea metalliferous sediments [6]. Mizuno and Moritani [7] described the

Fig. 2 Free fall grab for
manganese nodule
sampling, 1987

Fig. 3 Box corer with
manganese nodules at the
sediment surface, 1987

application of 3.5 kHz profiling in manganese nodule exploration, noting a relationship between manganese nodule abundance on the sea floor and the thickness of the
upper sediment layer as revealed by 3.5 kHz PDR profiling. With rare exceptions,
the thinner the upper sediment layer the greater the nodule abundance at the sediment surface, the greatest abundances of nodules occurring on the most slowly
accumulating sediments.

An important development in sea floor surveying in the late 1970s, which was massively developed during the 1980s and 1990s, was 'deep tow'. This was a platform housing a number of different instruments which could be towed near the sea floor thus considerably increasing resolution [8]. The system could include PDR, side scan sonar,3.5 kHz echo sounder, cameras, thermometers, and water sampling equipment. Seabeam was another important innovation that allowed for wide area survey and mapping at high resolution using 12 kHz for topography, 3.4 kHz for sub-bottom and 30 kHz for backscatter which helped to estimate nodule abundance (J Halkyard,pers. comm 2023).

Of the other geophysical methods of seafloor investigation, gravity never achieved any usage in marine mineral exploration, while magnetic and heat flow methods achieved a little use in the 1960s and 1970s, [9, 10]. Francis [11] reviewed the application of electrical methods of seafloor mineral exploration and concluded that their applicability was limited due to the high conductivity of seawater. Nevertheless, he considered that its potential, if any, lay in exploration for PMS deposits.

Towards the middle of the 1970s, marine minerals explorers started to integrate the various techniques available to them and by the late 1970s there was a tendency to move towards 'all purpose' mineral exploration systems [5]. Ships such as the Sonne (Germany) and the Hakuri Maru No 2 (Japan) were equipped with most if not all the latest marine mineral exploration devices and became the workhorses of the marine mineral exploration efforts of their respective countries in the 1980s and beyond.

Geochemical Methods

Unlike the deep-sea mineral exploration methods outlined above, geochemical exploration for deep-sea minerals could not, at least in the twentieth century, be carried out in real-time. For the most part, it relied on the chemical analysis of recovered material and the plotting of the data so obtained on maps and their interpretation. The formation of geochemical haloes in sediments around hydrothermal sulfide deposits in the Red Sea caused by dispersion and fall out of hydrothermal plume precipitates [12] has already been mentioned. The application of this technique to other areas, with some modification, came later. For example, it was applied to the TAG area on the Mid-Atlantic Ridge by Shearme et al. [13] who noted Fe Cu and Zn enrichments in sediments thought to have been introduced from hydrothermal vents as finely divided sulfides which were subsequently oxidized.

The geochemical dispersion techniques outlined in the previous paragraph were not applicable to the exploration for manganese nodules as nothing was dispersed away from them on the sea floor. The visual and remote techniques outlined above could only determine if nodules were present or absent on the seafloor. They could say nothing about their composition. Exploration for different compositional varieties of nodules had to rely on either routine grid sampling and chemical analysis, which was the method used overwhelmingly in the early days of nodule exploration, or un a thorough as possible understanding of the relationship between their composition and the environment of deposition of the nodules.

Mining Development

There was no commercial mining for deep-sea minerals in the twentieth century so it is only possible to write about what was proposed. The possibility of it occurring led to the development of several deep-sea mining systems. Two main things had to be taken into account in developing these systems First the great depth of the deposits to be recovered. Second, their occurrence for the most part widely distributed on the sea floor, and thus mining systems had to be able to cover large areas relatively quickly in order to achieve an economically viable rate of recovery.

Manganese Nodules

Smale-Adams and Jackson [14] reviewed early proposed methods for mining manganese nodules. Two systems were initially considered, a mechanical continuous line of buckets called the CLB system, and various types of hydraulic lift systems. The CLB system consisted of a continuous loop to which collecting buckets were attached. The buckets were supposed to descend to the sea floor in a conveyor-like manner, scoop up nodules and return to the surface. As originally conceived, the CLB system was to be operated from a single ship, but likely problems with this approach, including almost certain entanglement of the downward and upward moving parts of the system, led to the consideration of a two-ship system. According to Smale-Adams and Jackson, hydraulic lift systems consisted of a wide diameter pipe extending from the mining vessel to a manganese nodule collecting device. A thinner pipe ran part way down the main pipe and which could discharge either air, hydrocarbons, or light solid particles into it. Air was the medium overwhelmingly proposed. This produced a density contrast in the pipe which was designed to flush the nodules up it.

Details of early designs for nodule-collecting devices are few. This was perhaps the most proprietary aspect of deep-sea mining and companies were reluctant to publicize their designs. However, they all had to collect sufficient tonnages of nodules to maintain a high daily production rate, had to be able to recover a surface layer of nodules with minimum sediment disturbance or recovery, and be able to move easily over the sea floor without sinking in. Writing on the basis of Consortia submissions to the US National Oceanic and Atmospheric Agency (NOAA), Padan [15] reviewed proposed manganese nodule-collecting devices. The collector was to be either towed or self-propelled. Towed collectors would be pulled by the mining ship and their track was determined by the movements of the surface ship. Self-propelled collectors had the advantage of being able to follow a pre-determined path more easily. Smale-Adams and Jackson [14] described one version of a proposed manganese nodule collector being developed by the Kennecott Consortium. The nodules were to be collected by hydraulic suction possibly with the aid of nodule-dislodging devices. They considered that the number of nodules lifted by such a collector depended on the average abundance of the nodules, the width and forward speed of the collector, and its efficiency. The last of these depended in part on the width of the 'mouth' of the collector. Amman [16] gave some details of Preussag's proposed nodule collector. Nodules were to be picked up and separated from sediments on a forward-moving ramp and collected up prior to being lifted to the surface by hydraulic lift. The collector was of a tractor design [16].

In a presentation at the 17th Law of the Sea Institute [17] Jean Pierre Lenoble of AFERNOD outlined some of the technological problems that the French deep-sea mining program had been working on up to that date. Partly in order to avoid obstacles in the path of nodule collectors, AFERNOD had researched a "free swimming shuttle "concept for nodule mining. Using these, both obstacles and nodule-poor areas could be avoided. Such a shuttle, of about 1500 tonnes displacement, could dive from the surface with a weight on it as ballast. On its arrival at the bottom, by the discharge of part of its ballast and with the help of thrusters, it would crawl along the bottom and collect nodules. When the shuttle was full, the remaining ballast would be discharged, giving sufficient buoyancy for a return to the surface. It was envisaged that about 10–20 shuttles would operate around a mining ship which would be open at its stern with an internal harbour to handle the shuttles.

Scale model mining tests involving complete manganese nodule mining systems were conducted initially by the Deep Sea Ventures consortium (OMA) in 1970, and by the other consortia later in the 1970s, and were designed to operate at about one-fifth scale, about 1000 tonnes of nodules lifted per day. The tests were generally regarded as having been successful, although whether or not they could be scaled up to a full economic production of at least 5000 tonnes per day (10,000 tonnes according to Halkyard,pers. comm 2023) was never clearly established. According to Halkyard [18], Kennecott began its exploration for nodules in 1962 by dredging 10 tonnes of them off the coast of Baja California. From then until 1977 it had 13 more cruises, mostly in the CCZ. Smale-Adams and Jackson [14] described a test carried out in late 1974 and early 1975 of a scaled-down 7-tonne Kennecott hydraulic nodule collector towed with a steel armored electromechanical cable. It was designed to be operated at depths between 4000 m and 6000 m and towed at a velocity of approximately 2 knots.

Lipton et al. [19] have written on the nodule mining tests of the 1970s in the CCZ, which includes personal communications to John Parianos, one of its authors, from some of the industry participants. According to Lipton et al., based on personal communications from Kenecott's John Halkyard and David Felix, once their exploration was completed the Kennecott Consortium developed mining system designs, and numerical models and tested various components of a commercial mining system. The development work included a towed nodule pick-up system which was tested at approximately 1/5th scale, a hydraulic and airlift system which was tested on land, and study of nodule attrition, material handling, and transportation. The tests described by Smale-Adams and Jackson [14] (above) were also based on Kennecott's work.

Also according to Lipton et al. [19], based in part on personal communications from OMA's Tom Detweiler, in 1976 OMA started to trial mine nodules in the CCZ. OMA equipped the Wesser Ore, a 20,000-ton iron ore carrier, with a moon pool, a derrick, and revolving thrusters. The nodules were collected by a suction dredge towed on skis behind a rigid riser and were raised by airlift. The collector was at 1/5th scale except for the collector inlets which were at full design scale. The ship, renamed Deepsea Miner II, conducted its first tests in 1977 at 1900 km southwest of San Diego, California. Early in 1978, two further trials encountered difficulties when the dredge floundered in bottom sediment and a cyclone struck. The latter situation was dangerous as the riser could not be retrieved in time (it took well over 6 h to recover), so the ship, with the riser attached, had to ride out the the storm (later voyages included an emergency explosive detachment system for the riser). Finally, in October 1978, 550 tonnes of nodules were lifted in 18 h, at a maximum capacity of 50 tonnes/h.

Additionally according to Lipton et al. [19], based in part on personal communications from INCO/OMI's Ted Brockett, the International Nickel Company (INCO) made the decision to develop their collector systems in-house and set about designing what became internally known as the "Electro-Hydraulic" (EH) collector. INCO constructed a version of the EH collector designed to be tested on a cable and scheduled a deep-sea collector test in the CCZ in the early 1970s. Unfortunately, the collector, its instrumentation system, and a 7600-meter electro-mechanical tow cable were lost during a shallow water test off the coast of Oahu, Hawaii. In 1975 the OMI consortium was created, led by INCO. Early in 1976, INCO tested eight collectors. Two collectors were chosen for a subsequent pilot mining test and were constructed, one with a two-meter wide active collection width and one with a three-meter wide collection width. While the primary function of the collector was to gather the nodules from the seafloor, there were secondary functions; rejecting the oversized and undersized nodules, eliminating unwanted sediment, and introducing the nodules into the riser system. During the pilot mining testing, OMI tested both a hydraulic submersible pump and an air injection lift system to raise the nodules from the seafloor to the surface. In addition to providing a conduit for lifting the nodules to the surface, the riser pipe was used to tow and navigate the seafloor collector through the mine site. Once on deck, an air, water, and nodule separator directed the nodules to conveyors for transporting them to the ship's holds and on-deck storage containers. The OMI team, operating aboard the SEDCO 445 drill ship successfully recovered over 800 metric tonnes of nodules during these tests during the summer of 1978. The submersible pump system recovered approximately 650 tonnes of nodules while the airlift system recovered 150 tonnes. Nodule throughput varied dramatically throughout the tests with the rate exceeding 40 tonnes per hour at times causing the material handling systems and storage containers on the mining ship to overflow with nodules.

Finally, according to Lipton et al. [19], Ocean Minerals Company, in 1978, following the end of Project Azorian (see Chapter "Activities on Manganese Nodules

During the Post-war Boom") rented the Glomar Explorer from the United States Government to use as its pilot mining vessel. The OMCO partners thus sought to leverage off much of the already completed engineering work. Initial experiments were at a depth of 1800 m off California, but the first full tests were south of Hawaii at the end of 1978. These first tests were suspended because the doors of the moon pool refused to open. Finally, in February 1979, the operation was carried out with more success. OMCO's Charles Morgan [20] also described the Lockheed consortium (OMCO) 1978–1979 test in the CCZ of their prototype nodule mining system. The Lockheed collector was tested on the sea floor at a depth of around 4700 m. In addition to the collector (miner), the lowering comprised a buffer, load-bearing cable, hydraulic lines, and airlift system. According to Padan [15] one important conclusion from the test was that, within reason, sea state would not interfere with nodule mining.

In the presentation by J P Lenoble at the 17th Law of the Sea Institute [17] mentioned above, the transfer of nodules from mining ship to transport vessels and the latter's passage to a shore-based processing plant was discussed. He pointed out that transport distances from the most likely French mining site(s) would have been in the order of at least 3500–4000 km, depending on the location of the processing plant, on seas with wave heights of less than 4 m for 90% of the time, and wave periods of less than 10 s. These constraints would have affected the design of both mining and transport vessels. As daily nodule recovery was expected to have been between 3000 and 10,000 tonnes of dried nodules and the storage capacity of the mining ship between 30,000 and 100,000 tonnes, the number and size of ore carriers had to take account of this. It was expected that the nodules would have been transferred from the mining ship to the ore carriers by hydraulic slurry pumping, as was already being used for bulk transport of iron ore. This would have necessitated a preliminary grinding of the nodules to obtain a grain size of less than 1 mm. Based on these and other factors, AFERNOD calculated that between 2 and 8 ore carriers would have been needed, each of between 75,000 and 200,000 tons, depending on port capacity, in order to maintain the hoped for minimum production of 1.5 million tons of nodules per year.

Hydrothermal Deposits

In the 1960s and 1970s, the only hydrothermal deposits that were considered for mining were those in the Red Sea. According to Mustafa and Amman [21], mining of Red Sea muds, and by extension other unconsolidated or poorly consolidated hydrothermal deposits, could be done by suction after disaggregating the deposits using a vibration/cutting head, with seawater jetting and dilution to help mobilize the material up a steel pipe. A test mining operation in the mid-1970s demonstrated the feasibility of this process and also showed that the ore could be concentrated by shipboard flotation. The mining of other hydrothermal deposits was not seriously considered until the present century [22].

Co Rich Crusts

The possible mining of Co-rich crusts presented more of a problem than the mining of either manganese nodules or metalliferous muds. Co-rich crusts are for the most part firmly attached to the sea floor and would need to be broken off before they could be recovered. John Halkyard (pers. comm. 2023) proposed a concept for this purpose in the 1980s based on a vehicle being towed over the crust deposit and breaking it up prior to lifting. Halkyard further commented (pers. comm. 2023) that separation of thin (less than 5 cm thick) crusts from their underlying substrate would have required either a very precise cutting tool or a very efficient post-recovery separation in order to provide the 100+ tons per hour of broken up crust material needed in order to have made the operation economic.

References

1. Bruneau L, Jerlov N, Koczy F (1953) Physical and chemical methods. In: Petterson H (ed) Reports of the Swedish deep-sea expedition, physics and chemistry, vol 3, pp 101–112
2. Menard HW, Shipek CJ (1958) Surface concentrations of manganese nodules. Nature 182:1156–1158
3. Laughton AS (1967) The geology of the North Atlantic Ocean. In: Hersey JB (ed) Deep-sea photography. Johns Hopkins University, pp 191–206
4. Kaufman R, Siapno WD (1972, January 20–22) Variability of Pacific Ocean manganese nodule deposits. In: Horn DR (ed) Papers from a conference on ferromanganese deposits on the ocean floor. Arden House/Lamont-Doherty Geological Observatory/Office for the IDOE/NSF, Washington DC
5. Cronan DS (1980) Underwater minerals. Academic Press, 362pp
6. Backer H, Schoell M (1972) New deeps with brines and metalliferous sediments in the Red Sea. Nature 240:153–158
7. Mizuno A, Moritani T (1976) Annotated record of the detailed examination of Mn deposits recovered by the GSJ during the 1971-1974 period in the Pacific Ocean. In: Glasby GP, Katz HR (eds) Marine geological investigations in the southwest Pacific and adjacent areas, UNESCAP Tech Bull, vol 2, pp 62–79
8. Speiss F N, Lowenstein et al (1976) Fine-scale mapping near the deep-sea floor. In Oceans 76, Marine Technology Society, Washington D C
9. Weiss RF, Lonsdale P et al (1977) Hydrothermal plumes in the Galapagos rift. Nature 267:600–603
10. Rona P (1978) Criteria for recognition of hydrothermal mineral deposits in oceanic crust. Econ Geol 73:135–170
11. Francis TG (1987) Electrical methods in the exploration of seafloor mineral deposits. In: Teleki P et al (eds) Marine minerals advances in research and resource assessment, NATO ASI series Math and Phys Sci, vol 194. Plenium, pp 413–419
12. Bignell RD, Cronan DS, Tooms JS (1976) Metal dispersion in the Red Sea as an aid to marine geochemical exploration. Trans Inst Min Metall B 85:273–278
13. Shearme S, Cronan DS, Rona PA (1983) Geochemistry of sediments from the TAG hydrothermal field, MAR at latitude 26 N. Mar Geol 51:269–291
14. Smale- Adams KB, Jackson GO (1978) Manganese nodule mining. In: Sea floor development: moving into deeper water, Philos Trans Royal Soc Ser A, vol 290, pp 125–132

15. Padan JW (1990) Commercial recovery of deep seabed manganese nodules: twenty years of accomplishments. Mar Min 9:87–103
16. Amann H (1982) Technological trends in ocean mining. Philos Trans R Soc Lond Ser A 307(1499):377–403
17. Lenoble JP (1982) Problems and solutions. Paper presented at the 17th Law of the Sea Institute
18. Halkyard J (2022, October) Oral presentation. In: Underwater Mining Conference,St Petersburg, Florida
19. Lipton I, Nimmo M, Parianos J (2016) TOML Clarion-Clipperton Zone project, Pacific Ocean. NI43–101 report. AMC Consultants. Brisbane. www.sedar.com
20. Morgan C (2022, October) Oral presentation. In: Underwater Mining Conference,St Petersburg, Florida
21. Mustaffa Z, Amman H (1978) In Proceedings of the10th offshore technology conference, Houston, pp 1199–1206
22. www.solwaramining.org

Part II
Transition, Circumspection and Diversification, Early 1980s–2000

Introduction

The last two decades of the twentieth century were a period of considerable change in deep-sea minerals activities. There was a major transition in both perception and action after the conclusion of the Law of the Sea Conference in 1982, the Convention that came out of which being largely regarded as unfavorable by the embryonic marine mining industry. Economic factors also accelerated the downturn in manganese nodule-related work, a downturn which had commenced in the late 1970s. Nevertheless, the first part of the 1980s was a busy period in deep-sea minerals activities, before the implications of the Law of the Sea Convention had been fully evaluated and the economic downturn had not impacted the global minerals industry as much as it subsequently did.

Manganese nodules were the only minerals considered during the Law of the Sea Conference. During most of its deliberations, cobalt-rich crusts and hydrothermal deposits were so poorly known that they were ignored. Thus, Law of the Sea issues had little effect on activities related to those minerals, other than perhaps to encourage them as their development would not be significantly affected by the Convention and, as was subsequently realized, because they occur to a much greater extent than manganese nodules in EEZs.

Although by the early 1980s, manganese nodule-related activities by the US Consortia in the CCZ were running down (they conducted no new nodule exploration cruises after 1982), the work was taken up by State agencies. These also commenced nodule exploration in the Indian Ocean and expanded nodule exploration from the CCZ into the southwestern Pacific Ocean. Exploration for cobalt-rich crusts and hydrothermal deposits also increased in the Pacific, albeit by academic and government agencies rather than industrial concerns, and mainly in EEZs. There was certainly a perceived need for new sources of metals because at the time the USA depended completely on the import of manganese and cobalt and according to the Congressional Research Service in 1982 the Federal Government considered both to be among the six most strategic metals for the national economy. A study sponsored by the Congressional Budget Office evaluated the vulnerability of the US to politically motivated changes in price and availability of cobalt. The study concluded that Federally-funded research efforts to find new sources of Co should be supported.

In a well-timed article in 1983 [1], J Robert Moore summarised the then 'state of play' and attempted to predict the future as far as deep-sea minerals were concerned. He considered that manganese nodules would be disregarded in favor of hydrothermal deposits, perhaps until the end of the century. He also believed that interest in Co-rich crusts might further develop. He posed the question, why have we come to a period of no growth in nodule-related activities, at least by the industrial consortia in the CCZ? He suggested that it was because the international scene had been grossly misjudged, while a global economic recession compounded the difficulties inherent in nodule mining development. On a more optimistic note however, Moore pointed out that at the same time as the decline in the fortunes of nodules in the newly declared International Seabed Area was the discovery of PMS in the newly declared Exclusive Economic Zones, and that we were just at the beginning of research on the potential of those deposits. Looking ahead to the year 2000, Moore commented that having weathered the promotional enthusiasm of the 1960s and the economic difficulties of the 70s and early 80s, there was much to see that was still positive about deep-sea minerals. Assuming that the technological and institutional problems associated with marine mining could be solved, he thought that there would be some "good times" ahead in the marine mining business. Moore posed the question, "where in the World would future marine minerals activities take us"? He suggested Pacific spreading centers for PMS, and the Hawaiian Archipelago, US Trust territories, and the EEZs of Pacific Island states for Co-rich crusts. Nodules were already being sought in the last of these. Moore believed that for the USA to rise to the challenge of developing the oceans for marine minerals, firstly the finest efforts of its scientists and engineers would be needed, and secondly, its Government would have to encourage development with appropriate legislation, good regulation and commitment. During the remainder of the century,the US got the former but not the latter.

During the last decade of the twentieth century, deep-sea mineral activities were thus largely different from those that had taken place before. The emphasis on manganese nodule development declined, even amongst the state-supported consortia which had carried on after interest by the industrial consortia started to wane. Interest in Co-rich crust and PMS exploration increased, although not by industrial companies. Environmental concerns which had either been non-existent or thought to be easily manageable in the 1960s and 70s became of greater importance in the 1990s, resulting in a number of environmentally motivated programs and research cruises. Finally, generally static or declining metal prices throughout the period, coupled with the generally negative reaction from industry to the Law of the Sea Convention acted as a damper. During this period the transition away from the period of initial enthusiasm was completed and deep-sea minerals entered a period of circumspection and diversification.

Reference

1. Moore JR (1983) Marine hard minerals resources-progress and problems. Proc Oceans 83(111):1145–1149

Hydrothermal PMS and Related Deposits

Spreading Centres

According to Scott [1], in the 7 years following the discovery of the Galapagos hydrothermal field in 1979, actively forming and inactive seafloor PMS deposits containing
variable amounts of Fe, Zn, Cu, Ag, and Au had been found at more than 15 locations
associated with the mid-ocean ridge system in the eastern Pacific Ocean. Hydrothermal
deposits were also found in the Atlantic Ocean during this period, and in the western
Pacific [2] (Fig. 1). Some examples of the eastern Pacific deposits are given below.

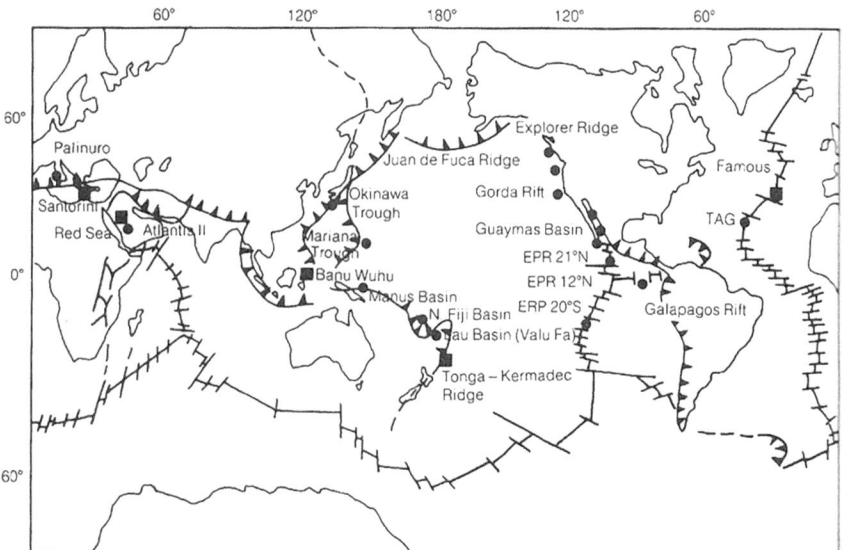

Fig. 1 Known worldwide submarine hydrothermal occurrences in the late 1980s

Some sites were found to have individual large deposits; others had numerous small scattered ones. Various combinations of sulphides, oxides and silicates were found. All were in axial or near axial locations on the ridges or located on seamounts near the ridge crests. All locations had high heat flow and likely sub-sea floor circulation of heated seawater, both thought to be prerequisites for PMS formation.

Scott (1987) commented on the size of the PMS deposits recovered up to the mid-1980s relative to that of terrestrial sulphide deposits. The largest PMS deposits discovered up until that time were those on the Galapagos Ridge, on the EPR at 13 N and on the Explorer Ridge in the Canadian EEZ. The Galapagos deposit was originally claimed to contain several million tonnes of ore [4] but this estimate was subsequently scaled down. The EPR at 13 N deposits were estimated to contain from two to four million tonnes of ore. The Explorer Ridge deposit was estimated to contain three to five million tonnes of ore in aggregate. Compared to these amounts, the largest terrestrial deposit to which sea floor PMS deposits were thought to be related, that in Cyprus, contains about 17.5 million tonnes of sulfides, putting the perceived 1980s economic value of the seafloor deposits into perspective. However,it should be noted that none of the early seafloor estimates were based on drilling.

Koski et al. [5] further described the sulphides mentioned in Part I from the Juan de Fuca Ridge, those collected in 1981 in the axial valley of the southern part of the ridge. Two types of massive sulphide were dredged, angular slabs of dark grey Zn rich sulphide with interlayers and a thin partly oxidised crust of Fe sulphide, and sub-rounded spongey fragments composed almost entirely of pale Fe -poor colloform sphalerite and opaline silica. Also the PMS that were found on the EPR near 13 N in 1981 were further described by Hekinian and Fouquet [6]. More than 80 active and inactive hydrothermal deposits were found, forming localised discontinuous fields averaging about 50 m in diameter in a band less than 200 m in width along a 20 km section of the axial graben. Each field was found to consist of irregular and conical shaped edifices varying from 1 m up to 25 m in height. They were made up of Fe, Cu and Zn sulphides. However, the most extensive hydrothermal deposits were found on the summit and flank of an off-axis volcano. These were about 800 m long and 200 m wide with a volume about ten times greater than those in the axial graben. This hydrothermal material was found to consist mainly of goethite overlying Fe rich massive sulphide, silica rich sulphide and massive Fe-Cu sulphide.

It was not only in the Pacific Ocean that exploration for hydrothermal minerals took place in the 1980s, but in the Atlantic and Indian Oceans too. In the Atlantic Ocean, the TAG hydrothermal field in the North Atlantic (Fig. 1) where PMS were discovered in 1985 after a history of exploration there since the early 1970s, is the best known and has been described by Rona [7] and references therein. In the Indian Ocean exploration for hydrothermal deposits commenced much later. In 1983 the German GEMINO project (Geothermal Metallogenesis Indian Ocean) was initiated to locate sites of hydrothermal activity on the Central Indian Ocean Ridge (CIOR) which found PMS in 1994 [8]. The first two GEMINO cruses in 1983 and 1986 did not find PMS deposits but did find other indications of hydrothermal activity on the CIOR, including hydrothermally altered basalts and hydrothermal metalliferous sediments. At the 1989 UMI, Herzig and Pluger(1989) [9] reported on the finding on Leg 1 of the GEMINO 3 cruise in 1987/88 of a volcanic ridge in the rift valley of a 30 km segment of the Ridge at 22 55 S which was a site of weak hydrothermal activity. On Leg 2 in December 1987 further evidence of hydrothermal activity was observed on the CIOR including an inactive field of hydrothermal deposits close to 23 S including dark grey blocks of supposed sulphide minerals up to 30 cm in diameter. The area was named the Sonne Hydrothermal Field.

Building on the work done on the GEMINO cruises, at the UMI in 1994 Halbach reported on massive sulphides recovered from the Sonne Hydrothermal Field [8] during the SO 92 cruise in 1994. The deposits were situated about 270 km NW of the Rodrigues Triple Junction southeast of Madagascar [8] and were the first recovered massive sulphides in the Indian Ocean. The sulphides were found at 23 23.5 S,69 14.45 E at a depth of 2840 m and covered an area of 300 400 m. The deposit was inactive at the time of sampling and the hydrothermal activity that formed it was thought, on the basis of its sediment cover, to have ceased at least several hundred years ago. The main sulphides recovered were pyrite, marcasite, chalcopyrite and sphalerite. They contained values of Cu up to 33% and Au up to 5 ppm [8].

Convergent Plate Boundaries

As outlined in Chapter "Hydrothermal Deposits: Discovery and Preliminary Economic Evaluation", from the mid-1960s up to the early 1980s the main efforts in deep-sea PMS exploration had been in the Red Sea and on mid-ocean Ridges in the eastern Pacific. Work on hydrothermal deposits in the western Pacific commenced in earnest in the mid-1980s.

Submarine hydrothermal deposits in the CCOP/SOPAC area of the SW Pacific received more attention from the point of view of their possibly being mined than any other such deposits outside of the Red Sea. Some of the richest PMS deposits so far found on the sea floor have been found in the SW Pacific. Reflecting this development, a report on the future of ocean research (FORE Report) prepared by the Scientific Committee on Ocean Research (SCOR) in the early 1980s said "It is in island arc and oceanic fracture zones that the best developed and most economically valuable hydrothermal sulphide deposits may be found". CCOP/SOPAC maintained a programme of exploration for hydrothermal deposits throughout the 1980s largely through bilateral arrangements with mostly French and German entities and through the ANZUS programme. The latter provided for research vessels supported jointly by Australia, New Zealand and the USA to work in the SW Pacific under aid arrangements for the CCOP/SOPAC member nations. Hydrothermal minerals investigations under this program commenced off Vanuatu in 1982 [10]. CCOP/SOPAC also benefited from cruises investigating hydrothermal deposits in its area from the late 1980s and throughout the 1990s by Japan under the SOPAC/MMAJ program which diversified away from manganese nodules (see below) into cobalt-rich crusts and hydrothermal deposits at the end of the 1980s. The EEZs of Fiji, Tonga, Vanuatu the Solomon Islands and PNG were the main targets, concentrating on the Bismarck Sea, the North Fiji Basin and the Lau Basin (Figure 2 of "Hydrothermal Deposits: Discovery and Preliminary Economic Evaluation"). Resource evaluation carried out by CCOP/SOPAC indicated that SW Pacific PMS deposits have high Cu, Au and Ag concentrations compared with the deposits globally, occur mostly close to land and in water depths not exceeding 2000 m on average.

The Bismarck Sea is divided into the New Guinea Basin to the west and the Manus Basin to the east. Some early work in the Manus Basin was carried out under the ANZUS program. Sea floor photographs taken there showed inactive chimney like structures [11] indicating past hydrothermal activity. Additional information on hydrothermal deposits in the Manus Basin showing was given by Scott and Binns [12].

The North Fiji Basin (NFB) is a 2000–4000 m deep back-arc basin standing between Vanuatu to the west and the Fiji Platform to the east (Figure 2 of "Hydrothermal Deposits: Discovery and Preliminary Economic Evaluation"). The first major research cruise to investigate hydrothermal activity in the NFB was by the Sonne (SO-35) between December 1984 and February 1985 [13]. Hydrothermal minerals were found consisting of pyrite associated with iron oxide coatings on rocks and yellowish- brown aggregates thought to be nontronite. Later, a scientific submersible dive in the NFB using the Nautile in 1989, found a PMS deposit of about 1 km diameter and 40 m thick [14]. It was situated in the axial graben. This was one of the largest hydrothermal deposits found on the sea floor outside the Red Sea up to that time.

The Lau Basin is between Tonga and Fiji (Figure 2 of "Hydrothermal Deposits: Discovery and Preliminary Economic Evaluation"). Two areas of hydrothermal activity were initially identified there, one in the northern Lau Basin on a feature named the Northern Lau Spreading Ridge (NLSR) (also called the Central Lau Spreading Centre by some authorities) and the second further south on a feature named the Valu Fa Ridge [13]. Porous hydrothermal nontronite was recovered from this latter region, together with ferromanganese oxide crusts. Subsequent work on Sonne cruise 48 in 1987 located more hydrothermal deposits on the NLSR including a hydrothermal chimney [15]. On the northern Valu Fa Ridge, Von Stackelberg et al. [15] observed, pyrite impregnated rocks, old sulphide chimneys and recovered some massive sulphide consisting mainly of sphalerite, pyrite, chalcopyrite and galena. Also on the Valu Fa Ridge, in 1989 a joint Franco/German scientific diving expedition using the Nautile discovered a large PMS deposit in the vicinity of 21.20 S [16].

According to Anon [17], work by the Geological Survey of Japan initiated Japanese work on evaluating submarine hydrothermal metalliferous deposits. Shipboard surveys were scheduled from 1985. The first of these was actually not at a convergent plate margin, but near Mexico where hydrothermal sulfides were recovered. However, most subsequent Japanese work on PMS and related deposits was carried out in the NW Pacific. This work was reviewed by Usui and Iizasa [18] during an Ocean Mining Symposium held at Tsukuba, Japan in November 1995.

The first report of PMS in a back-arc basin in the NW Pacific was made by Peter Halbach at the 1989 UMI and subsequently published in Economic Geology [19]. On June 26th 1988, the Jade Hydrothermal Field was discovered in the Okinawa Trough by the RV Sonne (Fig. 1). This cruise had taken place within the scope of a German-Japanese cooperative project. The hydrothermal field was found in a caldera like structure at 27.15 N and 127. 04.5 E The hydrothermal vent area was discovered in the northern part of the inside NE slope of the caldera in a water depth between 1300 and 1500 m. Various types of hydrothermal mineralization were identified, barite-rich sinter like material, a stockwork mineralization, and a sulfide-bearing sediment layer. The main sulfide minerals were found to be sphalerite, silver-bearing galena and pyrite. Some chalcopyrite was also found.

The mineralogical and geochemical investigations and the evaluation of the ship-born data resulted in the recognition that the Jade deposit is a modern analogue of the volcanogenic Kuroko type deposits which represent an economically significant subtype of the volcanogenic massive sulfide occurrences. These ore deposits had been exploited in Japan for centuries. The felsic country rocks of the Jade deposit are typical for a continental back-arc basin with a volcanogenic bimodal series of white rhyolite to dacite. The massive ore samples are rich in Zn and Pb and obviously represent fragments of chimneys which finally accumulated to form a hydrothermal mound deposit. Another kind of mineralisation was also identified and corresponds to the stockwork type in which fissures and cracks of a rhyolitic breccia are filled with pyrite and chalcopyrite [19].

According to Usui and Iizasa [18], submarine hydrothermal activity producing PMS and related deposits was subsequently found in the NW Pacific associated with submarine volcanoes in both island arcs and back-arc basins. The deposits occur as chimneys, mounds of broken down chimney material and disseminated sulphides in sediments and volcanic rocks. The deposits were found to be comparable in size to those on mid-ocean ridges. The most favoured locations for economic interest were thought to be active back-arc rifts in the Okinawa Trough and Mariana Trough and modern submarine volcanos and rifts on the volcanic ridges in the region.

End of Century PMS Activities

During the last part of the twentieth century, the largest number of new PMS discoveries were in the Pacific. Many of these were on the EPR, and in the EEZs of Canada, Mexico and the USA. Others were in the arcs and back-arc basins of the western Pacific which had been relatively neglected until the mid-late 1980s. Additional deposits were found on the Mid-Atlantic and Central Indian Ocean Ridges. Of the 139 seafloor hydrothermal sites tabulated by Rona and Scott [20] three-quarters were in the Pacific with about one-third of these in island arcs or other terranes.

At the 1992 UMI, Scott and Binns [21] reported on the discovery in 1991 during the PACMANUS Expedition (Papua New Guinea-Australia-Canada in the Manus

Basin) of a very large (approx 800 × 350 m) actively forming PMS deposit in the southern part of the Manus Basin (Fig. 1) in water depths of 1650–1700 m. Videos of the deposit showed large spires, chimneys and mounds with abundant fauna. Two small samples from an active chimney were recovered, containing chalcopyrite, bornite, tennantite and sphalerite. Gold values for two samples were 2 and 10 ppm. Scott and Binns [12] provided further information on this discovery and classified it as a PMS similar to the Jade and Valu Fa deposits. Rona [22] updated information on the PACMANUS deposit and noted that in 1997 the Nautilus Minerals Corporation leased from the PNG Government two sites in the PACMANUS hydrothermal field in order to evaluate it for mining. This work identified the SOLWARA deposit and continued for over 20 years [23]. A timeline of the PACMANUS investigations and their salient findings has been given by CCOP/SOPAC [24].

In the North Fiji Basin work continued on PMS deposits throughout the 1990s. The French -Japanese STARMAR Project of the late 1980s and early 1990s discovered an active hydrothermal mound named "White Lady" and an inactive "Pere Lachaise" site [25]. In 1995, west and NW of Pere Lachaise a new hydrothermal field was discovered on SO cruise 99 between 16.58 S and 16.57 S [26]. It was observed to be an elongated field about 500 × 800 m and was found to contain numerous chimneys capping hydrothermal mounds up to 10 m high. At the 2000 UMI,Halbach reported on new PMS investigations done in 1998 on cruise SO134 in the North Fiji Basin which found chimneys younger than those found on the SO 99 cruise [27]. The principal sulfide minerals observed included chalcopyrite, cubanite, bornite, covellite, pyrite, marcasite, sphalerite and wurzite. In 1999, as part of the MMAJ/SOPAC Programme, the Hakuri Maru No 2 carried out a survey in the NFB [24] re-examining previously found PMS deposits and looking for new ones. Temperature anomalies were detected at 22 locations and evidence of sulphides at 13 locations. Massive sulphide ore was collected in the northern part of the axial valley.

After the discovery of hydrothermal manganese oxide crusts on the Tonga-Kermadec Ridge in 1981 [28], a search for PMS deposits both on the TKR and on other arcs in the SW Pacific was carried out. In 1998, Wright et al. [29] reported on the discovery of hydrothermal sulfide mineralization from southern Kermadec Ridge (Fig. 1) volcanoes, Brothers and Rumble II West. Sulfides recovered from the two volcanoes included chalcopyrite, sphalerite, marcasite, galena, and pyrite and contain up to 15.3% Cu and 18.8% Zn. These discoveries provided the first direct evidence of the long predicted high-temperature hydrothermal activity on the Kermadec Arc and stimulated additional research both there and in other Pacific arcs.

Studies on hydrothermal PMS deposits in the Atlantic and Indian Oceans continued during the last part of the twentieth century. German work in the Indian Ocean has been reviewed above [8]. Several hydrothermal fields had been found on the MAR (Fig. 1) of which the TAG hydrothermal field was the best known. At the 1992 UMI, Rona [30] reported on collaborative dives with the submersible Alvin in 1991 and with two Russian MIR submersibles in 1992 down to the TAG hydrothermal field. They found and sampled inactive sulfide mounds as well as the active mound discovered in 1985. The active mound was found to be about 250 m in diameter and

40 m high with a concentric zonation of mineral types and flow regimes decreasing in temperature from 365 degrees black smokers at the center to lower temperature white smokers and diffuse flow towards the margins. Inactive sulfide chimneys there were found to contain the first primary free gold grains found at any hydrothermal site on a mid-ocean ridge [31]. Work on additional hydrothermal sites on the MAR and the deposits found in the twentieth century was reviewed by Cherkashov [32].

The marine exploration activities in the southern part of the North Atlantic Ocean also resulted in the discovery of a further type of marine hydrothermal mineralization: ultramafic-hosted PMS deposits [33]. Ultramafic-hosted rocks are commonly exposed on portions of slow-spreading ridges near transform faults. The reaction of hydrothermal seawater with ultramafic rocks produces low pH values and much higher levels of methane than does the reaction of seawater with Mid Ocean Ridge Basalt [33]. These are strongly indicative of serpentinization processes which may also may produce some heat flux. On the Mid-Atlantic ridge from 15° N (Logatchev site) to 36° N (Rainbow site) there exists significant out-cropping of ultramafic rocks. It was thought possible that Rainbow and Logatchev are hybrid MORB-ultramafic systems. The PMS formed in these ultramafic-hosted hydrothermal systems are on average very rich in Cu and in Au (Cu: 17.9%; Au:10 g/t;) [34]. The high Cu content is mainly caused by the presence of chalcopyrite and isocubanite, which are typical high-temperature minerals.

Many additional hydrothermal sites were found on the Mid-Atlantic Ridge, and elsewhere, in the twenty-first century (see Epilogue).

References

1. Scott SD (1987) Seafloor polymetallic sulfides: Scientific curiosities or mines of the future. In Teleki PG et al (ed) Marine minerals advances in research and resource assessment. NATO ASI Series Ser C 94.D Riedel, p 277–300
2. Cronan DS (1985) Marine mineral resources. Geol Today 1:115–118
3. Shearme S, Cronan DS, Rona PA (1983) Geochemistry of sediments from the TAG hydrothermal field, MAR at latitude 26 N. Mar Geol 51:269–291
4. Malahoff A (1982) A comparison of the massive submarine polymetallic sulfides of the Galapagos Rift with some continental deposits. Mar Tech Soc J 16(3):39–45
5. Koski RA, Normark WR, Reid JA (1984) Tectonic and volcanic controls on hydrothermal processes in the southern Juan de Fuca Ridge area. Geol Soc Am Bull 95:603–615
6. Hekinian R, Fouquet Y (1985) Volcanism and metallogenesis of axial and off-axial structures on the East Pacific Rise near 13 N. Econ Geol 80:221–249
7. Rona PA, Klinkhammer TA et al (1986) Black smokers, massive sulphides and vent biota at the mid-Atlantic Ridge. Nature 321:33–37
8. Halbach PE, Blum N, Pluger W (1994) The Sonne Ore field-The first massive sulfides from the Indian Ocean floor. Underwater Mining Institute
9. Herzig P, Pluger W (1989) Discovery of hydrothermal fields at the Central Indian Ridge (GEMINIO project). Underwater Mining Institute
10. Exon N, Cronan DS (1983) Hydrothermal iron deposits and associated sediments from submarine volcanoes off Vanuatu, Southwest Pacific. Mar Geol 52:43–52

11. Both R, Crook K et al (1986) Hydrothermal chimneys and associated fauna in the Manus back-arc basin. Papua New Guinea. EOS Trans Am Geophys Union 67:489–491

12. Scott SD, Binns RA (1995) Hydrothermal processes and contrasting styles of mineralization in the western Woodlark and eastern Manus Basins of the western Pacific. In: Parson LM et al (eds) Hydrothermal vents and processes, Geological Society London Special Publication 87, pp 191–205

13. Von Stackelberg and the shipboard scientific party (1985) Hydrothermal sulfide deposits in back-arc spreading centres in the southwest Pacific. B.G.R. Circular No 2, B.G.R. Hannover, 14pp

14. Auzende JM, Urabe T et al (1989) Preliminary results of the STARMER 1 cruise of the submersible Nautile in the North Fiji Basin. Abs. Joint CCOP/SOPAC-IOC Fourth International Workshop on Geology, Geophysics and Mineral Resources of the South Pacific. CCOP/SOPAC Suva Fiji, p 12–13

15. Von Stackelberg and the shipboard scientific party (1988) Active hydrothermalism in the Lau back-arc basin (SW Pacific)-first results of the Sonne 48 cruise (1987). Mar Min 7:1–14

16. Fouquet Y, Stackelberg UV et al (1991) Hydrothermal activity and metallogenesis in the Lau back-arc basin. Nature 349:778–781

17. Anon (1985) Study on Submarine Hydrothermalism in the Izu-Ogasawara Arc Area (project, FY 1984-1988) Geological Survey of Japan

18. Usui A, Iizasa K (1995) Deep -sea mineral resources in the Northwest Pacific Ocean: geology, geochemistry,origin and exploration. Proc ISOPE Ocean mining symposium (1995) Tsukuba, Japan, p 131–137

19. Halbach PE, Pracejus B, Maerten A (1993) Geology and mineralogy of massive sulfide ores from the Central Okinawa Trough, Japan. Econ Geol 88:2210–2225

20. Rona PA, Scott SD (1993) Preface. In seafloor hydrothermal mineralization: new perspectives. Econ Geol 881:193–1976

21. Scott SD, Binns RA et al (1992) PACMANUS: an actively forming submarine polymetallic sulfide deposit in felsic volcanic rocks of the Manus back-arc, Papua New Guinea. Underwater Mining Institute

22. Rona PA (2008) The changing vision of marine minerals. Ore Geol Rev 33(3-4):618–666

23. www.solwaramining

24. Tawake A (Unpublished) Summary Report on marine minerals exploration in the Pacific Islands Region. SOPAC Secretariate, Suva, Fiji

25. Bendel V, Fouquet Y et al (1993) The white lady hydrothermal field, North Fiji back-arc basin, Southwest Pacific. Econ Geol 88:2237–2247

26. Halbach P, Auzende M et al (1995) HYFIFLUX cruise: German-French cooperation for the study of hydrothermalism and related tectonism, magmatism and biology of active ridges in the North Fiji Basin (SW Pacific). InterRidge News 4:37–43

27. Halbach P (2000) The modern massive sulfide deposits in the North Fiji Basin (NFB): Results from the SO 134 cruise in August/September 1998. Underwater Mining Institute

28. Cronan DS, Glasby GP, Knedler K et al (1982) A submarine hydrothermal manganese deposit from the southwestern Pacific Island arc. Nature 298:456–458

29. Wright IC, de Ronde CEJ et al (1998) Discovery of hydrothermal sulfide mineralization from southern Kermadec arc volcanoes (SW Pacific). Earth Planetry Sci Lett 164:335–343

30. Rona PA (1992) Seafloor hydrothermal mineralization: new discovery. Underwater Mining Institute

31. Rona PA et al (1991) Eos Trans (American Geophysical Union) 72:470–471

32. Cherkashov GA (1995) Hydrothermal input to sediments of the Mid-Atlantic Ridge. In: Parson LM et al (ed) Hydrothermal vents and processes. Geological Society London Special Publication 87, p 223–230

33. Wetzel LR, Shock EL (2000) Distinguishing ultramafic -from basalt-hosted submarine hydrothermal systems by comparing calculated vent fluid compositions. J Geophys Res 105:8319–8340

34. Cherkashov G (2017) Seafloor massive sulfide deposits: distribution and prospecting. In: Sharma R (ed) Deep-Sea mining. Springer International Publishing AG, Cham, pp 143–164

Expanding Cobalt-Rich Crust Activities in the Pacific Ocean

Introduction

Cobalt-rich crust studies blossomed after the conclusion of the Law of the Sea Conference in 1982 and the establishment of the right of coastal states to declare an EEZ. Crusts did not occur only in EEZs but they were mostly sought there. Hein et al. [1] outlined the history of some of these activities in the Central Pacific based on 10 German and US cruises after the 1981 Sonne 18 cruise outlined in Chapter "Cobalt Rich Crusts: Recognition and Preliminary Evaluations", Usui and Someya [2] outlined Japanese activities in the NW Pacific, and CCOP/SOPAC coordinated a number of crust activities in the EEZs of its island member states in the SW Pacific in the 1980s and 90s [3]. Cobalt-rich crust studies in the Indian and Atlantic Oceans were far fewer than in the Pacific during this period, mainly due to the much more limited distribution of the deposits in those oceans, especially in the Atlantic. Roonwal [4] provided an outline of what little Co-rich crust information there was from the Indian Ocean.

North Pacific

Following the Sonne 18 cruise in the North Pacific (Chapter "Cobalt Rich Crusts: Recognition and Preliminary Evaluations"), further FRG crust cruises took place in 1984 (SO 33),1985 (SO37) and 1986 (SO46). Peter Halbach was Chief Scientist. According to Hein et al. [1] results from these and contemporary studies confirmed the salient features of the crusts outlined in Part I of this book. The crusts coat exposed rock substrates preferentially between about 1000 and 2500 m depth (Fig. 1) and contain up to about 2% Co. They exhibit decreasing Co with increasing depth, Pt is enriched in some crusts, there are two growth generations in some crusts

© The Author(s), under exclusive license to Springer Nature Switzerland AG 2024
D. S. Cronan, *Deep-Sea Minerals Developments in the 20th Century*,
https://doi.org/10.1007/978-3-031-52342-7_8

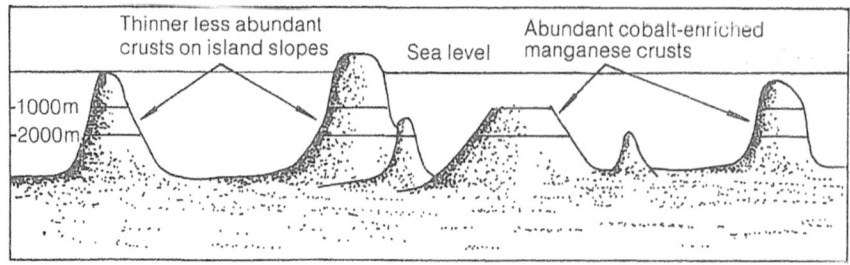

Fig. 1 Settings of Co-rich crusts on seamounts, guyots and island slopes

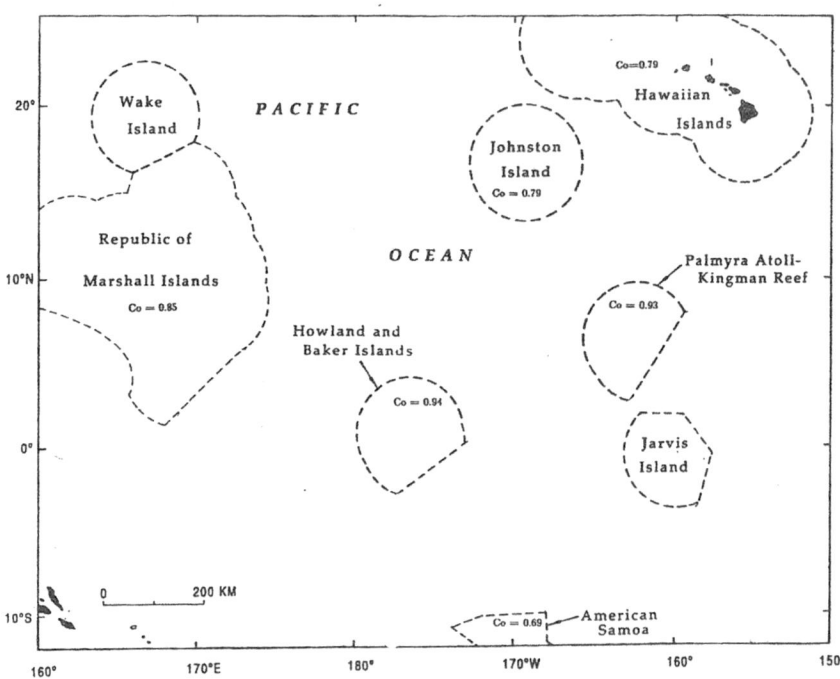

Fig. 2 Locations of Co-rich crust studies in US Pacific territories up to circa 1990, with average Co values of the crusts obtained in each territory (reference in text)

separated by a layer of phosphorite, and most (75%) crust growth rates vary between <1 and 5 mm\my [5].

The US also carried out crust studies in the Pacific in the 1980s, (Fig. 2) [1] mainly by the USGS in the EEZs of the US Territories there, but the University of Hawaii also got involved. The US activities in this regard were bolstered by President Regan's declaration of a 200 nautical mile EEZ around the USA and its overseas territories and were established within the USGS Office of Energy and Marine Geology. Coordination of the programme was from the Pacific Branch of Marine

Geology. The first USGS cruise in the study of Pacific crusts was in 1983 aboard the SP Lee. J R Hein and F. Manheim were co-chief scientists. Two FRG representatives participated. This cruise was the first attempt by the USGS to study crusts on seamounts in US territories in the northern Line Islands. Four USGS scientists participated in the 1984 Sonne cruise, which took place in the Hawaii area and also returned to the northern Line Islands. Also in 1984, there was a short cruise on the SP Lee to study crusts in the EEZ of the US Trust Territory of the Marshall Islands and which sampled crusts from seven seamounts, and the University of Hawaii mounted two cruises to study crusts in the Hawaiian EEZ [6]. Another University of Hawaii cruise was made in 1987 to study crusts on the Manihiki Plateau, where generally thin crusts were found, and in the central Line Islands [7]. The main purposes of all these cruises were to (i) to refine understanding of both the local and regional compositional variations in Co-rich crusts, and their variation in relation to depth, (ii) determine the oceanographic, geological and biological processes that influenced the nature and distribution of the crusts, and (iii) determine the resource potential of the crusts not only for Co but for other elements of possible resource interest such as Ti, Ni, Pb, Mo Ru and Pt. Of these latter elements, Pt and Ru engendered quite a lot of interest as they are rare precious metals with limited terrestrial sources, commonly in ultrabasic rocks. Platinum concentrations in the Pacific crusts investigated were found to range from about 0.1–0.9 ppm and tended to co-vary with Ni in the deposits [8]. These values are more than 80 times the Pt values of the earth's crust and more than is found in manganese nodules. Ruthenium concentration in a crust from the Mid-Pacific Mountains was found to be 3.80 parts per billion [9]. Ultrabasic rocks contain from about 0.70–2.0 ppb Ru.

The Japanese started to show an interest in Co-rich crusts in the mid 1980s and studies on them by MMAJ and the Japanese University sector commenced in 1985 [10]. In 1986, a committee report by the Japanese Agency of Natural Resources and Energy recommended that there be an "immediate commencement of exploration" for the deposits [10]. In 1987, MITI commenced at sea surveys as part of a 10 year programme in response to the above recommendation [10]. The main focus of Japanese crust research was in the NW Pacific. According to Usui and Someya [2], the thickest crusts, more than 10 cm thick, were recovered from NW Pacific seamounts and plateaux mostly older than 70 million years in water depths shallower than 2500 m. Deposits were found to contain up to 1.5% Co and variably high contents of Ni, Pb, Y, V and Pt. Manganese, Co and Pt were found to be enriched in deposits at water depths of between 1000 and 2500 m, with Ti and Ni showing sporadic enrichments in these depths. Usui and Someya [2] charted the distribution of Co-rich crusts in the NW Pacific and concluded that crusts on seamounts and plateaux there had an economic potential.

Korean interest in crusts commenced in the late 1980s under the guidance of Jung-Keuk Kang at KORDI. Jointly funded USGS-KORDI cruises occurred in 1989 to the Marshall Islands EEZ, 1990 to FSM, and 1991 to Palau and FSM [11, 12]; KORDI funded cruises, with USGS participation occurred in 1997 and 1998 to the Marshall Islands EEZ (J Hein, pers. comm. 2023).

The first Co-rich crust investigations by China were in 1987,and systematic investigations began in 1997 [13]. The work included investigations in the Magellan Seamounts area and in the Mid-Pacific Mountains by COMRA (China Ocean Mineral Resources R &D Association).

South Pacific

Interest in their Co-rich crusts by the Island states of the South Pacific commenced in the mid-1980s,somewhat later than by Germany and the USA in the North Pacific. This was largely due to to their coordinating body CCOP/SOPAC not having the capability to recover crusts, as it did not have access to a vessel large enough to deploy the heavy dredges needed for crust recovery at that time. Nevertheless, by the late 1980s, sufficient crust data had been collected in the CCOP/SOPAC region from various sources for compilation studies to be carried out. The first of these was by Hodkinson and Cronan [14] who used 333 crust analyses for Mn, Fe, Co, Ni and Cu and concluded that the two principal influences on crust composition in the region were water depth and latitude. Subsequent studies have been reviewed and updated by Mizell et al. [15] and Verlaan and Cronan [16].

Cobalt-rich crust investigations in the CCOP/SOPAC region got fully underway after the commencement of the Japan/SOPAC Cooperative Deep-Sea Mineral Resources Study Programme,1985–2005 [17]. This program encompassed the EEZs of all the CCOP/SOPAC states (Fig. 3). The work was carried out by the state-of-the-art research vessel Hakurei- Maru No 2 (Fig. 4). Cobalt-rich crust cruises

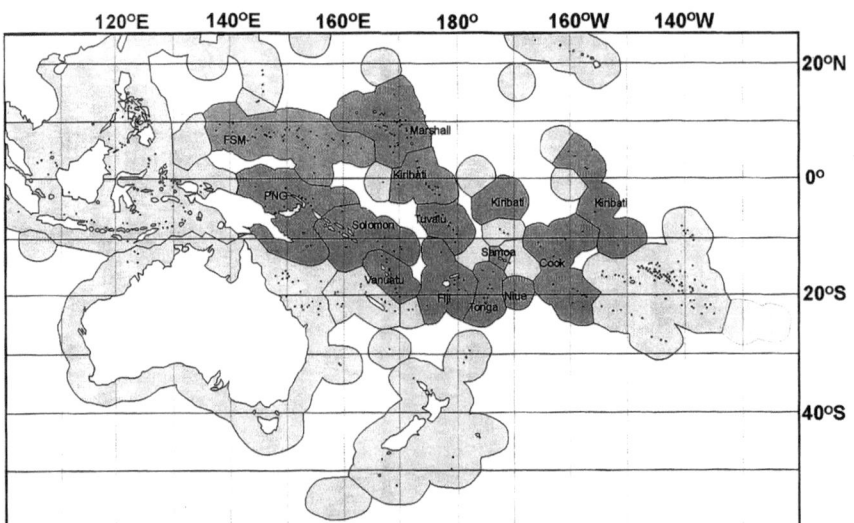

Fig. 3 Nations participating in the Japan-SOPAC program,1985–2000

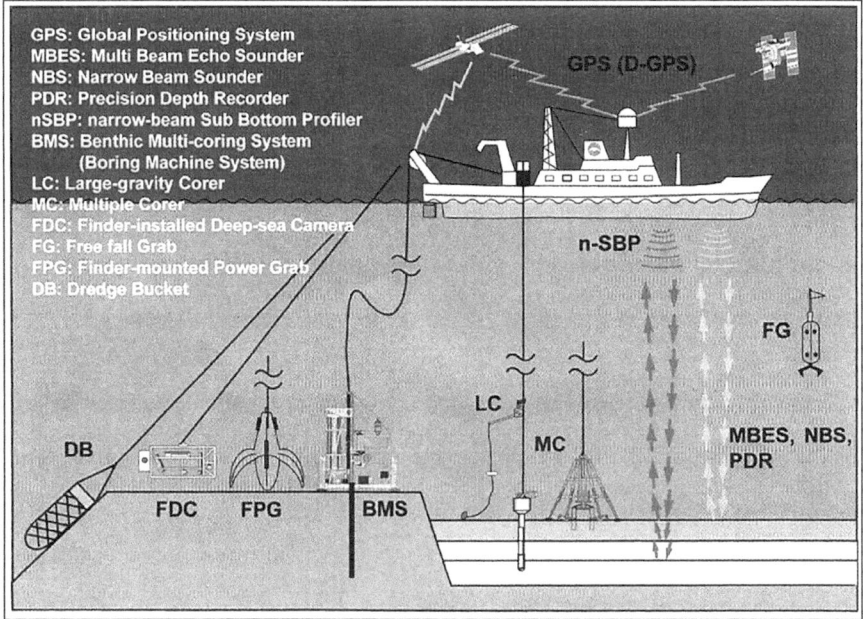

Fig. 4 Schematic of operational facilities on Hakurei-Maru No 2 (reference in text)

commenced in 1987 and continued until the end of the century, at first in conjunction with manganese nodule studies (see Chapter "Late 20th-Century Manganese Nodule Activities") and then, in the late 1990s, alone [17].The surveys to assess the potential of Co-rich crusts were conducted in the EEZs of Kiribati (Gilbert, Phoenix, and Line Islands), Tuvalu, and Samoa, as well as in the EEZs of the North Pacific CCOP/SOPAC associate member countries, the Marshall Islands and the Federated States of Micronesia. Crust thicknesses varied from under 10 mm to over 40 mm (Fig. 5). In the Gilbert Islands Co was found to average around 0.7%, in the Phoenix Islands around 0.8%, and in the Line Islands around 0.6% [18]. Detailed results of the Japan/ SOPAC cruises can be found in the annual reports submitted to SOPAC by the MMAJ [19].

Resource Potential

According to Hein et al. [1] the central Pacific was considered the prime area for any economic exploitation of metals in Co-rich crusts. Areas that received significant resource-based studies were the Hawaiian Islands, Johnson and Palmyra Islands, the US Trust and affiliated territories, [20], the Cook Islands area [21], and the Phoenix Islands area [22]. The metals of interest were primarily Co, with Mn, Ni and Pt of subsidiary interest. Hein et al. [23] outlined eight criteria that could be used in

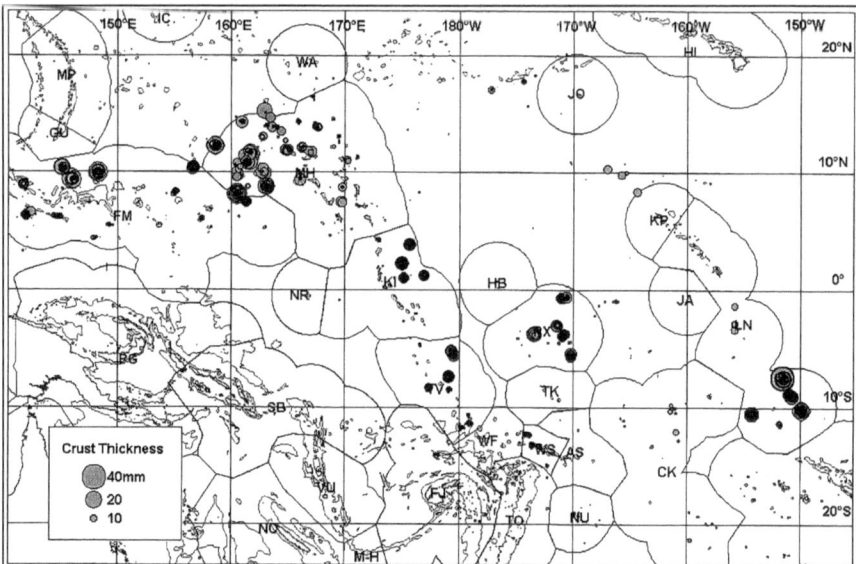

Fig. 5 Cobalt-rich crust thicknesses in the SOPAC region

exploration for potentially economic crusts and three criteria for their exploitation namely Co 0.8% and above, thickness 4 cm or greater, and gentle topography. The more important exploration criteria included, volcanic edifices shallower than 1500–1000 m, substrates older than 20 my, strong current activity, a shallow and well-developed oxygen minimum zone, stable slopes and no active volcanism.

The earliest investigations on Co-rich crusts specifically from a possible mining point of view were in the US EEZ in the Pacific. The first of these was on the Horizon and SP Lee seamounts south-west of Hawaii. According to Richey [24] an investigation of crusts on these seamounts was carried out aboard the SP Lee (L5-83-HW) in 1983 by the USGS with US Bureau of Mines participation. Systematic sampling and seismic reflection surveys were made on the seamounts together with camera and video surveys over portions of them. Bottom photographs showed extensive areas of crust outcrop. Maximum crust thickness was 5.0 cm on Horizon Guyot and 4.5 cm on SP Lee Guyot. An average crust thickness of 2.5 cm was used in resource calculations. According to Richey [24] estimated crust resources between 800 and 2400 m depth on Horizon Guyot were found to be about 75 million tonnes containing 18 mt of Mn, 1 mt of Ti, 0.58 mt of Co and 0.34 mt of Ni. On SP Lee Guyot within the same depth interval were estimated to be 24 mt of crusts containing about 7 mt of Mn, 0.32 mt of Ti, 0.28 mt of Co and 0.13 mt of Ni. Estimated Pt resources were about two million ounces in total. Richey [24] estimated that an annual production of 1 mt of crust would supply virtually all the then US Co needs and a significant portion of its needs for the other metals too and that this would continue for Co for the following 10 years even if only 10% of the crusts on the two seamounts were to be commercially exploitable.

In 1984, the US Authorities announced that the investigation of the Co-rich crusts occurring in the EEZ of Hawaii would be funded by an $1.8 million grant from the Minerals Management Service. The study was to centre on the assessment of crust resource potential and any possible environmental effects of mining. US Authorities sub-divided the EEZ surrounding the Hawaiian Islands into four geographic areas in regard to Co-rich crust occurrences [25]: A, Northwest Ridge Seamount Area, B, Central Seamounts, C, Central Ridge and D, Southeast seamounts. Based on this assessment, the US Minerals Management Service joined with the State of Hawaii Department of Planning and Economic Development to examine possible future crust mining in the Hawaiian EEZ. One outcome of this collaboration was the commissioning of an Environmental Impact Statement (EIS) which evaluated the possible impacts of future exploration, mining and processing of the deposits. This was of historical significance as it was the first EIS on Co-rich crusts anywhere. Impacts that were considered in the EIS included those associated with the selection of the mine site, the testing of mining equipment and the commercial mining of the deposits for a period of 20 years. According to its author, CL Morgan [26] "the major objective was to identify and examine the major issues of concern in mining and processing the crusts and to facilitate their resolution by outlining the data collection and research results which should be obtained during the exploration phase".

In 1985, the US Minerals Management Service published the results of an extensive study by Clark et al. [27] on Co-rich crusts in the EEZs of the US Trust and Affiliated Territories in the Pacific, drawing together much of the information that had been gathered earlier in the decade. A resource assessment was undertaken based on several criteria, including that:

1. Crusts of economic interest occur only on seamount or guyots older than 25my.
2. Their maximum thickness occurs within the depth range 800–2400 m.
3. Maximum crust coverage occurs within 5–15 degrees of the equator.
4. Presence of two generations of crusts,16-9my and < 8my.
5. Areas of greatest economic potential are where both crust generations are present, in other words, thick crusts.

According to P Halbach (pers. comm. 1984) areas with older generation crusts only might be economically viable depending on the thickness and composition of the crusts, but it was unlikely that any economically viable crusts could be found amongst the younger generation only.

From the above, the areas of greatest Co-rich crust potential were considered to be the Marshall Islands, Kingman-Palmyra, Johnston and Wake Islands. Those with lesser potential included, in order of decreasing potential, Federated States of Micronesia, Marianas, Jarvis, Belau-Palau, Guam, Howland-Baker and American Samoa. These were early conclusions, expected to change with the acquisition of additional data. This has proved to be the case. According to Hein (pers. comm. 2023), in terms of EEZs, the eastern part of the EEZ of the Commonwealth of the Northern Mariana Islands and the Republic of the Marshall Islands probably have the greatest crust potential. Additionally, the formation of Marine National

Monuments has limited the availability of Co-rich crusts, eg the Hawaiian Papahanaumokuakea Marine National Monument and Sanctuary; (https://www.papahanaumokuakea.gov). This is the largest protected marine conservation area in the world. The rest of the Hawaii EEZ may also become off-limits to deep-sea mining as,at the time of writing, the Hawaii State Government are debating a total ban on it in the Hawaiian EEZ.

In the South Pacific the main areas of activity for crusts from the resource point of view were the Cook Islands, and the Phoenix and Line Islands (Kiribati). Crust thicknesses in the Cook Islands area were found to range from 2 to 40 mm and average 14 mm. Maximum thicknesses were recorded between 1200 and 1500 m [21]. In the Phoenix Islands area, crust thicknesses were found to range from 1 to 50 mm and average 19 mm. Maximum thicknesses occur in the 2100–2300 depth range [22], deeper than in the Cook Islands area. In Line Islands crusts, Aplin [28] recorded an average of 0.55% Co in crusts collected by the University of Hawaii in 1979. In addition, De Carlo and Fraley [7] noted crusts of between 3 and 12 mm thick in the Line Islands EEZ at water depths below 2000 m, also collected by the University of Hawaii. As mentioned above, in the 1990s Co-rich crusts became an important focus for the Japan/SOPAC collaboration on marine minerals in the SW Pacific, many were collected by the Hakuri Maru No 2, and resource estimations based on them are described in the annual reports of that programme [19].

References

1. Hein JR, Schulz MS, Gein LM (1992) Central Pacific cobalt-rich ferromanganese crusts: historical perspective and regional variability. In: Keating BH, Bolton BR (eds) Geology and offshore mineral resources of the Central Pacific Basin. Circum-Pacific Council for Energy and Mineral Resources, Earth Science Series,14, Springer Verlag, p 261–284
2. Usui A, Someya M (1997) Distribution and composition of marine hydrogenetic and hydrothermal deposits in the northwest Pacific. In: Nicholson K, Hein JR et al (eds) Manganese mineralization: geochemistry and mineralogy of terrestrial and marine deposits. Geological Society, London, Special Publication 119, p 177–198
3. Tawake AK (2010) Summary report on marine minerals exploration in the Pacific Islands region, 1st draft, SPC-SOPAC, Suva, Fiji (unpublished)
4. Roonwall G (2017) Indian Ocean resources and technology. Taylor & Francis
5. Hein JR, Koschinsky A et al (2000) Cobalt-rich ferromanganese crusts in the Pacific. In: Cronan DS (ed) Handbook of marine mineral deposits. CRC Press, pp 239–279
6. Chave KE, Morgan CL, Green WJ (1986) A geochemical comparison of manganese oxide deposits of the Hawaiian archipelago and the deep sea. Appl Geochem 1:233–240
7. De Carlo EH, Fraley CM (1992) Chemistry and mineralogy of ferromanganese deposits from the equatorial Pacific Ocean. In: Keating BH, Bolton BR (eds) Geology and offshore mineral resources of the Central Pacific Basin. Circum-Pacific Council for Energy and Mineral Resources, Earth Science Series,14, Springer Verlag, p 225–246
8. Halbach P, Puteanus D, Manheim FT (1984) Platinum concentrations in ferromanganese seamount crusts from the Central Pacific. Naturwissenschaften 71:577–579
9. Bekov GI, Letokhov VN et al (1984) Ruthenium in the ocean. Nature 312:748–750
10. Anon (1988) Report of Japanese activities on cobalt-rich ferromanganese oxide crusts. Japan Agency of Natural Resources and Energy (unpublished)

11. Kang JK, Hong SY (1992) Current status of Korean deep seabed mining in Korea. Underwater Mining Institute

12. Hein J et al (1990) USGS Open File Report 90-407

13. Shi X, Bu W et al (2004) Co-rich ferromanganese crusts from the mid-and western Pacific, their characteristics,distributions and genesis based on the Chinese investigations. Conf. Minerals of the Ocean: Integrated Strategies-2. St Petersburg, 25–30th April 2004

14. Hodkinson RA, Cronan DS (1991) Regional and depth variability in the composition of cobalt-rich ferromanganese crusts from the SOPAC area and adjacent parts of the central equatorial Pacific. Mar Geol 98:437–441

15. Mizell K, Hein JR (2019) Geographic and oceanographic influences on ferromanganese crust composition along a Pacific Ocean meridional transect, 14 N to 14S. Geochem Geophys Geosyst. https://doi.org/10.1029/2019GC008716

16. Verlaan PA, Cronan DS (2022) Origin and variability of resource grade marine ferromanganese nodules and crusts in the Pacific Ocean; a review of biogeochemical and physical controls. Geochemistry 82:125741

17. Okamoto N (2006) Deep- Sea mineral potential in the South Pacific waters-the results of the 21 year long term Japan/SOPAC cooperative deep-sea mineral resources study programme. Underwater Mining Institute

18. SOPAC (n.d.) Deep-Sea minerals in the Pacific Islands region. Information Brochure 4, Kiribati. SPC-SOPAC, Suva, Fiji, p 12

19. Metal Mining Agency of Japan (1986–2000) Ocean Resource Investigations in the Sea Areas of CCOP/SOPAC. Annual Reports on the Joint Basic Study for the Development of Resources. Various Countries. SPC-SOPAC, Suva, Fiji

20. Johnson CJ, Clark AL, Otto JM (1985) Unpublished. In: Hein JR, Schulz MS, Gein LM (1992) Central Pacific cobalt-rich ferromanganese crusts: historical perspective and regional variability. In: Keating BH, Bolton BR (eds) Geology and offshore mineral resources of the Central Pacific Basin.Circum-Pacific Council for Energy and Mineral Resources, Earth Science Series,14, Springer Verlag, p 261–283

21. Cronan DS, Hodkinson RA et al (1991) An evaluation of manganese nodules and cobalt-rich crusts in South Pacific EEZs part 1, nodules and crusts in and adjacent to the EEZ of the Cook Islands. Mar Min 10:1–28

22. Cronan DS, Hodkinson RA (1991) An evaluation of manganese nodules and cobalt-rich crusts in South Pacific EEZs part 2 nodules and crusts in and adjacent to the EEZ of the Phoenix Islands. Mar Min 10:267–284

23. Hein JR, Schwab WC, Davis AS (1988) Cobalt and platinum rich ferromanganese rich crusts and associated substrate rocks from The Marshall Islands. Mar Geol 78:255–283

24. Richey JL (1987) Assessment of cobalt-rich manganese crust resources on Horizon and S P Lee Guyots, US EEZ. Mar Min 6:231–243

25. Oceans (1984) Geochemistry of ferromanganese crusts from the Hawaiian Archipelago – I. Cobalt-rich crusts. Proceedings of the Oceans 1984 Conference, p 421

26. Morgan CL (1990) Proposed marine mineral lease sale: Exclusive Economic Zone adjacent to Hawaii and Johnson Island. Final Environmental Impact Statement Volume I. U.S. Department of the Interior Minerals Management Service and the State of Hawaii Department of Business and Economic Development

27. Clark AL, Humphrey P et al (1985) Cobalt-rich manganese crust potential, OCS Study MMS 85–0006, U S Dept of the Interior, Minerals Management Service, p 35

28. Aplin A (1983) The geochemistry and environment of deposition of some ferromanganese oxide deposits from the south equatorial Pacific. Ph.D. Thesis, University of London, Applied Geochemistry Research Group, Imperial College

Late Twentieth-Century Manganese Nodule Activities

International Seabed Area

The worldwide distribution of manganese nodules, as perceived in the mid-1980s is shown in Fig. 1.

North Pacific

Industrial Consortia Activities

New manganese nodule exploration by the US consortia had largely ceased by the early 1980s and they had moved on to other nodule-related activities. According to the Financial Times and other sources the positions of the these Consortia immediately prior to the conclusion of the Law of the Sea Conference in 1982 were:

1. The Lockheed Group (OMCO) had retained about 30 staff to help lobby Government, work on legal matters, and keep a watching brief on polymetallic sulfides.
2. The Kennecott Group was concentrating on securing title to possible mining sites and advising the Government on Law of the Sea developments. Most of Kennecott's marine mining staff had been laid off but were still available on a consultancy basis. Other members of the Consortium only had staff part-time on the deep-sea mining issue.
3. The Inco/AMR Consortium (OMI) was processing data from earlier exploration, developing mining techniques, and working on production problems.
4. The OMA Group (Deep Sea Ventures) was continuing work on evaluation and production problems. According to Glasby [1] it was the last of the Industrial Consortia to withdraw from deep-sea mining evaluation.

© The Author(s), under exclusive license to Springer Nature Switzerland AG 2024   95
D. S. Cronan, *Deep-Sea Minerals Developments in the 20th Century*,
https://doi.org/10.1007/978-3-031-52342-7_9

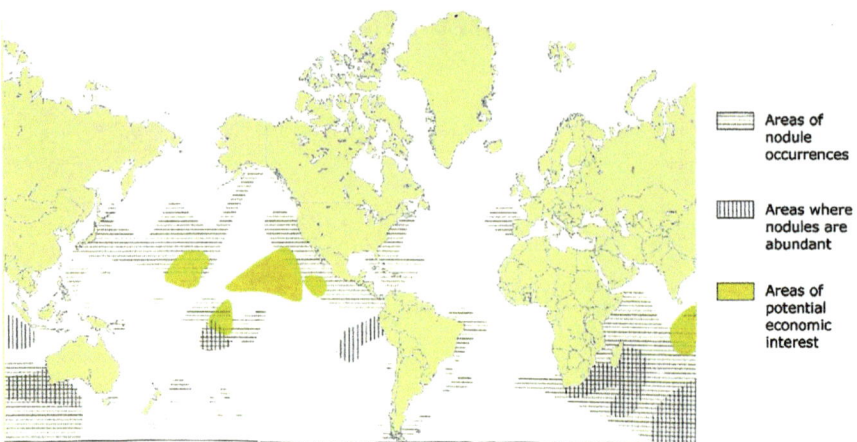

Fig. 1 Known distribution of manganese nodules in the oceans circa the mid-1980s

Information on the activities and distribution of claims by the Industrial manganese nodule mining Consortia in the mid-to-late 1980s has been well summarised by Earney [2], having been obtained in part by direct contact with the companies concerned. This latter approach was of considerable benefit as much, if not most, of the information on Consortia activities has never been put into the public domain. The composition of the Consortia in which the USA was involved 1987 was also given by Earney:

1. Ocean Management Inc. (OMI) comprised companies from the USA, Canada, FRG and Japan with INCO as its founder.
2. Ocean Minerals Company (OMCO) was entirely US-based with Lockheed and Cyprus Minerals having equal shares.
3. Ocean Mining Associates (OMA) comprised companies from the US, Belgium, and Italy with Deep Sea Ventures as its service contractor.
4. The Kennecott Consortium (KCON) comprised companies from the US, UK, Canada, and Japan.

The first detailed analysis of Consortia acquired data on nodule composition in the CCZ put into the public domain was by Knoop et al. [3]. Using data collected during 16 OMCO exploration cruises in the CCZ, Knoop et al. made a synthesis of nodule nature and variability and attempted to develop a quantitative regional scale assessment of the relationship between nodule composition and depositional environment. The data set used comprised 4657 grab samples collected from an area of approximately 1.5 million square kilometers, roughly 3000 × 500 km in size. The results indicated that diagenetic accretion was the predominant growth mechanism. This increased in relative importance with decreasing latitude and longitude from northwest to southeast. This was explained in terms of regional variations in primary productivity and sedimentation rate. Morgan [4] followed up this work with a resource assessment of the nodules based on a statistical analysis of the same data.

The area of the CCZ included in this analysis represents about 15% of the Pacific basin and was estimated to hold about 340,000,000,000 tonnes of nodules. At the then (1998) World production rate for Ni, Morgan concluded that the nodules in the area could provide more than 350 years' worth of production.

National Consortia Activities

The National Consortia were not so directly focused on profit as were the Industrial Consortia; security of supply was deemed sufficient reason to get involved in nodule mining. Although by the late 1980s "strategic metal" considerations were not so important as they had been earlier, security of supply was still seen as a valid reason to undertake what might initially seem not to be a profitable venture. Locations of the National Consortia claims are given in Fig. 2.

The French manganese nodule program had been functioning since 1973 under the AFERNOD Project. Up until 1980 the main emphasis was on nodule exploration, whereas post-1980 much effort was devoted to the development of mining systems, especially 'free swimming' shuttles (see Chapter "Technological Developments,1980–2000"). At the same time, chemical processes for extracting metals from nodules were defined, and simulations of economic conditions were undertaken, most of which indicated that nodule mining would not have been profitable at the then prevailing metal prices. Based on this early work, in 1983 it was decided to gather the expertise of all organizations involved in French nodule work into a G.I.P (Group of Public Interest) called GEMONOD. According to Hoffert [5], French industry became less and less interested in nodules during the 1980s, and the exploration of the area of the CCZ claimed by the French became less intense. However, in 1988, the submersible Nautile was used in the first-ever evaluation of nodules from a manned submersible. After that, according to Hoffert [5], the French nodule project "went into a period of hibernation", from which it has yet to emerge. Indeed, at the time of writing, France is one of the countries reported to be opposed to deep-sea mining.

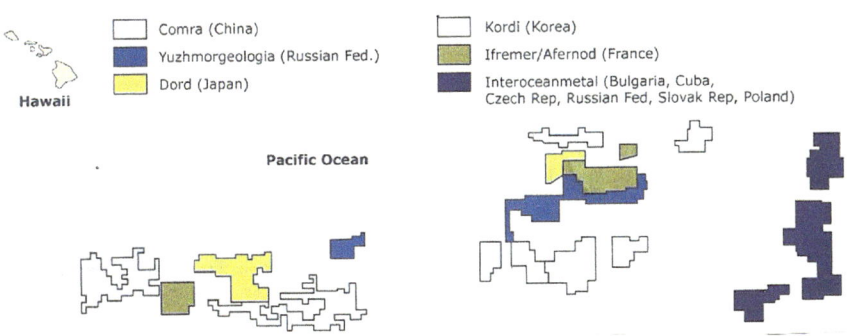

Fig. 2 National consortia exploration claims in the CCZ. (Source ISA)

USSR nodule studies were already well established in the CCZ by the 1980s, having commenced there in the 1960s. A considerable amount of additional data was collected by the Russian enterprise Yuzhmorgeologia during the period 1980–1990 [6] which was used to refine models to explain nodule variability in the Pioneer Investor area awarded to the Russians. Work to establish the causes of variations in nodule grade was undertaken by Korsakov et al. [7] which helped to develop a geological model for the Russian area. Later, Yubko and Gorelik [8] analysed and reviewed nodule distribution in the CCZ in relation to depositional environment, geological structure, and age.

Much of Germany's original work on nodules had been done by industrial concerns like Preussag AG and Metallgesellschaft AG which later joined the OMI (INCO) Industrial Consortium. However, they also maintained an independent role in the wholly German AMR Group. The German Geological Survey (BGR) was also heavily involved [9]. This work, up to the late 1980s, was extensively reviewed by Halbach et al. [10], including the nature and origin of the nodules in the German exploration areas, their environments of deposition and possible environmental problems in mining them, and their economic importance together with exploration, mining and processing considerations. At the time of writing, Germany still retains an interest in manganese nodule mining [11].

Japan had been involved in CCZ nodule work since the 1970s and after voting for the Law of the Sea Convention in 1982, continued to develop it. According to [12] in 1982 Japan passed a law to protect future Japanese nodule operations. This 'Deep Seabed Provisional Law' (1982) allowed the Japanese Government to establish companies jointly with the Private Sector. The first of these was DORD (Deep Ocean Resources Development Corp). Japanese policy on this matter has been reviewed in [13].

As well as maintaining an interest in the CCZ throughout the 1980s and beyond, Japan also continued to develop its interest in nodules in the Central Pacific Basin, the only nation to do so. Kohpina and Usui [14] attempted to estimate the resource potential of CPB nodules on the basis of more than 1000 analyses on nodules collected in the CPB. They calculated that around 9000 million tons of nodules with abundances greater than 5 kg per sq. m containing 39 mt of Cu, 52 mt of Ni and 27 mt of Co were expected to occur in 660,000 sq. km of the area.

Korea was not involved in nodule operations prior to the Law of the Sea Convention but after 1983 KORDI (Korea Ocean Research and Development Institute) started carrying out surveys on nodules in the CCZ [15] It was tasked with selecting optimal mining sites, and carrying out environmental studies to meet expected environmental regulations. The latter was an example of a possible future nodule mining entity carrying out environmental evaluation from the outset. Initially, the work was carried out using the RV Kana Keoki of the University of Hawaii. From 1989 to 1991 Korea conducted its nodule exploration programme using the RV Farnella, on charter, in collaboration with the USGS. In 1995, the Korean nodule programme was reorganised under the auspices of KADOM, a mixture of government and private entities. Korea, as a latecomer to nodule exploration, was not awarded its final allocated area by the ISA until 2002, and KORDI was confirmed as the main institute coordinating Korea's manganese nodule efforts into the future [15].

China was also not involved in manganese nodule operations prior to the Law of the Sea Convention. According to Lu Wenzheng [16], COMRA (China Ocean Mineral Resources R&D Association) commenced its exploration and assessment of nodules in the CCZ in the late 1980s. According to Jin [17] because China entered into nodule exploration after most of the other nations involved, it was difficult for it to select a high-quality working area. The area finally selected was at the western end of the CCZ and actually consisted of two areas 200 km apart with a 1500 km long spread from east to west. Nodules were found in abundances varying up to about 20 kg per sq. m and were more abundant in the west than in the east. By contrast, nodule grade (Ni + Cu + Co) was found to be higher in the east (2.86%) than in the west (2.2%).

The IOM Consortium initially comprised socialist countries of eastern Europe plus the USSR, Cuba and Vietnam, with its headquarters located in Poland. It was created in 1987. In 1989, Vietnam retired, East Germany retired in 1991. Later IOM activities are given in [18].

South Pacific

Much of the South Pacific comprises EEZs, particularly in its western part (Fig. 3). Nodules in EEZs will be dealt with later. However, in the period leading up to 1983 there were a number of wider-ranging studies on South Pacific nodules (see Part I) of which the last was by Exon [19] who published a synthesis of manganese nodule variability in the south-central and southwestern Pacific. The area considered was around 9 million sq. km, extending from the Marshall Islands to the Line Islands in

Fig. 3 World EEZs and International Seabed Area

the north and from Niue to the Austral Islands in the south. It included 380 stations, allowing a first assessment of the economic potential of its nodules to be made. Exon concentrated on four basins, the East Central Pacific, North Penrhyn, South Penrhyn and Southwestern Pacific Basins, and found nodules to occur at 80% of stations in the last two of these and at 70% in the first two. The South Penrhyn and Southwestern Pacific Basins were found to contain "low grade" nodules (that is low Ni + Cu + Co, maximum 2.02%) in high abundances of up to 62 kg per sq. m. The North Penrhyn Basin was found to contain nodules of "moderate grade" (maximum 2.10%, Ni + Cu + Co) in low abundances. The East Central Pacific Basin was found to contain "high grade" nodules of up to 3.55% (Ni + Cu + Co) in moderate abundances. Exon concluded that major nodule concentration and grade variations were largely controlled by plankton productivity and that the East Central Pacific Basin offered the greatest economic potential for them.

Based on the Aitutaki-Jarvis Transect of 1987 (see Part I), Cronan and Hodkinson [20] and Cronan [21] attempted to refine the conclusions of Exon [19] in regard to nodule variability in the southwestern Pacific and the controls on it. They found that superimposed on primary plankton productivity [19], additional controls on the composition of the nodules were the nature of their associated sediments and their distance above or below the CCD. Fluxes of Mn, Ni and Cu to the nodules were found to increase in equatorial regions of high biological productivity, as noted by Exon [19], but high values of these elements only occurred within about 200 m above or below the CCD. Under the most highly productive waters nearer the equator, Mn, Ni and Cu-rich nodules did not appear to occur at any depth, nor were the nodules abundant there. Also of considerable interest on the Aitutaki-Jarvis Transect was the confirmation of abundant deposits of Co-rich nodules in the central South Pacific. These occurred south of the equatorial zone of high productivity. Sometimes these deposits occurred in concentrations of more than 30 kg per sq. m and occasionally reached more than 60 kg per sq. m. They were found to contain up to 0.6% Co. Mero [22] had first drawn attention to Co-rich nodules in the central South Pacific, but prior to the 1980s they had not been thought to be of any economic interest. Subsequently, from the economic point, of view they became the most sought-after nodules in the region, especially when they were later found to be enriched in Ti, V and some Rare Earth Elements [23].

Indian Ocean

A map showing the main physiographic features of the Indian Ocean and manganese nodule sampling sites as of the early 1980s is given in Fig. 4.

By 1983, the Indian National Institute of Oceanography had surveyed for nodules in the Central Indian Ocean Basin down to 10 S, between 70 E and 90 E and found good grades but low overall abundances, generally not more than 5 kg per sq. m. Sometimes there were high abundances, but these were generally large nodules rich

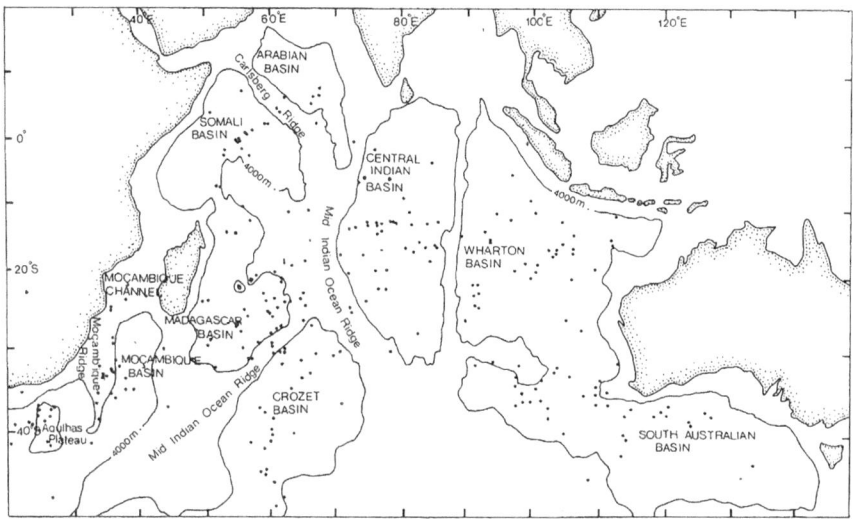

Fig. 4 Manganese nodule sample sites in the Indian Ocean up to the early 1980s. (reference in text). Note the much-increased sampling since the early 1970s (Figure 5 of "Activities on Manganese Nodules During the Post-war Boom") leading to the CIOB being found to contain nodules rich in Mn, Ni, Cu, and Zn

in iron and low in Ni, Cu and Co, and thus of little or no economic interest. Some E-W variation in the nodules was found. Small nodules were found to occur near the 90 E Ridge and larger nodules near the Chagos-Laccadive Ridge. By 1985, as a follow-up, the shipping company J Marr & Son of Hull UK, announced that its research vessel G A Reay would do further work on the CIOB nodules following the signing of a charter agreement with the Indian Department of Ocean Development [24]. This followed the successful completion in 1984 of a 120,000 km exploration program in the Indian Ocean by another of its ships, the Farnella.

Following on from the work done in the CIOB with charter ships in the early 1980s, the Indians employed the newly acquired R V Saga Kanya, obtained from West Germany, to further their subsequent work on nodules there. A summary of the activities during this period was given by Sudhakar [25]. Two main areas were studied in the Central Indian Basin [25]. These activities resulted in India being granted a "Pioneer Area" by the Preparatory Commission of the ISA in 1987. More than 95% of the area allocated was between 10 and 15 S on mostly siliceous sediment. The average Ni + Cu + Co was 2.25%, similar to grades reported in the CCZ. Shama [26] and Jauhari and Pattan [27] also provided an account of Indian Ocean manganese nodule investigations during this period and later, and tabulated the achievements.

Exclusive Economic Zones

In the post UNCLOS III period, the right to declare EEZs of 200 nm (more under specific circumstances) engendered an upsurge in marine minerals-related activities off coastal states.

Blake Plateau

The first EEZ nodules to be of interest were those on the Blake Plateau off the SE coast of the USA (see Part I). These had been known for some time but the declaration of an EEZ around the USA by President Regan in 1983 raised new interest in them as they would be wholly within that EEZ. The US Dept of the Interior announced that it would offer mining leases for portions of the Blake Plateau as early as 1983. The Blake Plateau nodules tended to be poorer in Ni and Cu than CCZ nodules but some were richer in Co. The interest in them at the time was based on their shallow depth, considerable abundance, and proximity to the USA. As well as their metals, their possible use for automobile catalyst materials was an additional factor in maintaining interest in them (see Chapter "Activities on Manganese Nodules During the Post-war Boom"). Until the emergence of the EEZs off the NW USA, Hawaii, and the U S Trust Territories in the Central Pacific as possible sources of deep-sea minerals in the mid-1980s, the Blake Plateau offered the largest potential source of marine minerals within the jurisdiction of the United States, but no mining has taken place there up to the time of writing.

SW Pacific

CCOP/SOPAC had been carrying out manganese nodule exploration cruises in the areas around its member nations since the 1970s (see Part I). In 1984, the Metal Mining Agency of Japan (MMAJ) agreed to carry out manganese nodule exploration and survey cruises within the EEZs (declared or to be declared) of CCOP/SOPAC island nations over the following 5 years under the Japan/SOPAC project (see Chapter "Expanding Cobalt-Rich Crust Activities in the Pacific Ocean"). The cruises were to be one each year by the Hakurei-Maru No 2 of approximately a month to 6 weeks in duration. The EEZs of the Cook Islands, Kiribati, Samoa, and Tuvalu were targeted, commencing with the Cook Islands in 1985.This programme was subsequently widened to include other deep-sea minerals and other areas and extended into the twenty-first century. However, nodules were the first minerals to be studied in the EEZs of all the islands where they were thought to occur in sufficient quantities to be of economic interest (Fig. 5). The last MMAJ/SOPAC nodule cruise in the Cook Islands was in 2000. The detailed results of this work were published in a

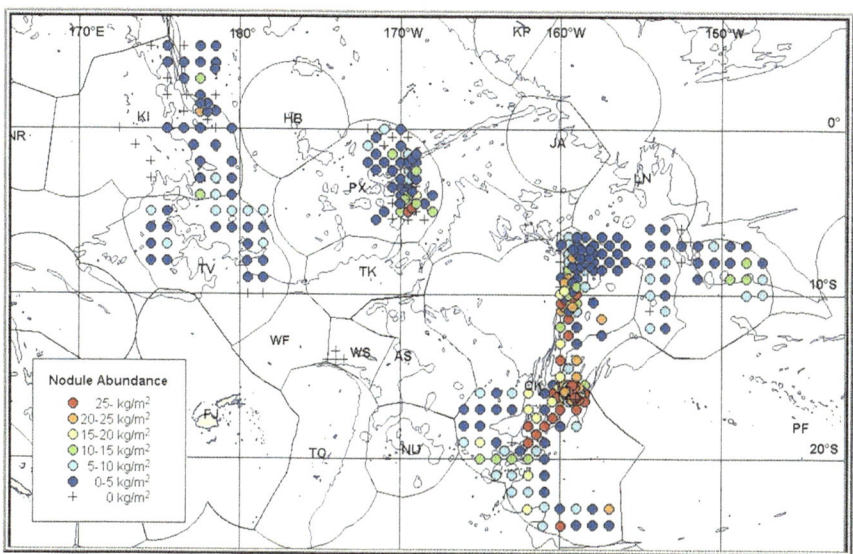

Fig. 5 Manganese nodule abundances in SOPAC member countries

series of volumes under the general title 'Ocean Resources Investigations in the Sea Area of CCOP/SOPAC. Report on the Joint Basic Study for the Development of Resources', produced annually in the year following the relevant cruise [28]. They are the prime source of information on deep-sea minerals in the SOPAC area. Summary results of these and various other studies on manganese nodules in the EEZs of island countries in the SW Pacific in the 1980s were outlined by Cronan et al. [29] and Kinoshita and Tiffin [30], and were also plotted in the MMAJ/SOPAC South Pacific Sea Floor Atlas [31]. In the Cook and Line Islands (Kiribati) EEZs the highest nodule abundances were found to occur between about 10–15 S with maximum Co occurring in the same general area, but maximum Ni of between 1.25% and 1.7% and Cu of between 1.25% and 1.4% occurring mostly within 5 degrees of the equator. In the Phoenix Islands EEZ (Kiribati), greatest nodule abundances of over 25 kg per sq. m were found to occur in a small area around 5 S,170 W in the SE of the EEZ and greatest Ni and Cu of about 1.5% each near the equator. Cobalt was found to be highest (0.4–0.5%) in the Phoenix Islands EEZ around 4 S.

In addition to the Hakuri Maru No 2 programme, in 1986 two cruises of HMNZS Tui were undertaken by the Australian and New Zealand Governments as part of the CCOP/SOPAC Tripartite Program to assess the abundance, metal content and distribution of marine minerals in parts of the SW Pacific that had not been well surveyed at that time. The Tripartite Program was an aid program carried out on CCOP/SOPAC's behalf by the USA, Australia and New Zealand. Part of this work involved the investigation of nodules in the Cook Islands EEZ east and west of the Manihiki Plateau [32]. Those to the east were found to be richer in Ni and Cu than those to the west. Also as part of the Tripartite Program there was a cruise of the RV Moana Wave of the University of Hawaii to the Phoenix Islands in 1986 [33]. Nodule abundances were found to range from 7 to 34 kg per sq. m. The combined Ni + Cu + Co content of the nodules averaged only 1.7% A few samples contained more than 2% combined Ni and Cu but were present in abundances of less than 5 kg per sq. m, the generally accepted cut-off grade at the time for manganese nodule mining. Moana Wave also worked in the Line Islands EEZ and collected nodules from seamounts in the Line Islands chain at about 2 S containing average Co of 0.67% [34] which showed affinities with Co-rich crusts in the region.

As well as studies at sea on manganese nodules in the EEZs of the SW Pacific during the final two decades of the twentieth century, a number of compilation and synthesis studies, some with an economic bias, were also carried out. At that time, the Scripps Manganese Nodule Data Base (see Part I) which was started in the 1970s was in widespread use. Mc Kelvey et al. [35] wrote a commentary on the Scripps Data Base in which they drew attention to an area between 124 and 160 W, 0-18 S which they believed could be a target for nodules containing more than 1%combined Ni + Cu. Further,they drew attention to the area between 154 and 160 W,1–3 S which they thought could be an exploration target for even richer nodules containing more than 1.8% Ni + Cu. In 1995, the East-West Centre, Hawaii, produced a report [36], which compared the Cook Islands EEZ with the CCZ and concluded that further resource oriented work in the former was justified. Following on shortly was a report by the Bechtel Corporation of San Francisco on a Deep-Sea Nodule Mining Prefeasibility Study [37]. This was an economic assessment and concluded, given a number of assumptions, that a nodule mining project in the Cook Islands EEZ would be feasible so long as the price of Co did not drop below $US 16.75 per pound in April 1996 dollars. Finally, soon after the Bechtel work was a Norwegian study [38] which attempted among other things to refine the economic conclusions of the Bechtel study. Conducting a sensitivity analysis by varying the price of Co, the capital costs and operating costs of a CI nodule mining project, the Norwegian study concluded that the project economics were more sensitive to changes in the price of Co than to reasonable changes in capital and operating costs. At the then price of Co of around $US 15 per pound the study concluded that manganese nodule mining for Co in the CI EEZ was not feasible at that time and should await higher Co prices. No Cook Islands nodule mining project has taken place, but work on them from the economic point of view by the Cook Islands Seabed Minerals Authority is continuing at the time of writing (J Parianos, Pers. Comm. 2023).

References

1. Glasby GP (1986) Marine minerals in the Pacific. Oceanog Mar Biol Ann Rev 24:11–64
2. Earney FCF (1990) Marine mineral resources. Routledge, p 387
3. Knoop PA, Owen RM, Morgan CL (1998) Regional variability in ferromanganese nodule composition: northeastern tropical Pacific Ocean. Mar Geol 147:1–12

4. Morgan CL (2000) Resource estimates of the Clarion-Clipperton manganese nodule deposits. In: Cronan DS (ed) Handbook of marine mineral deposits. CRC Press, pp 145–170
5. Hoffert (2008) Les Nodules Polymetalliques. Societe Geologique de France. Vuibert, p 431
6. Kasmin Y (2009) Geology of the Clarion-Clipperton fracture zone: existing geological information in respect of polymetallic nodules in the CCZ. In: Establishment of a geological model of polymetallic nodule deposits in the Clarion -Clipperton fracture zone of the equatorial north Pacific. International Seabed Authority, Kingston, Jamaica, p 106–144
7. Korsakov OD (Ed) (1987) Formation, environment and distribution trends of ferromanganese nodules of the World Ocean. Nauka, Moscow, p 287
8. Yubko VM, Gorelik IM (1992) Age estimates of manganese nodules on the basis of geological data. In: Geology of Ocean and Seas, Abstracts of the 10th international school on marine geology, 3. Russian Academy of Science and P Shirshov Institute of Oceanology, p 107
9. von Stackelberg U, Beiersdorf H (1987) Manganese nodules and sediments in the equatorial North Pacific Ocean "Sonne Cruise SO 25,1982" Geologisches Jahrbuch, Reihe, Heft 87, p 403
10. Halbach P,Friedrich G, von Stackelberg U (Eds) (1988) The manganese nodule belt of the Pacific Ocean. D Enke Verlag, Stuttgart, p 254
11. www.bgr.bund.de Manganese nodule exploration in the German license area
12. Anon (1982) Min J
13. www.taylorfrancis.com Developing a manganese nodule policy for Japan
14. Kohpina P, Usui A (1996) Estimation of manganese nodule resources in the northern part of the Central Pacific Basin. Bull Geol Survey Japan 47:255–271
15. Park C-K, Kim K-H et al (2003) Exploration strategies and activities for deep-sea mineral resources development of Korea. Underwater Mining Institute
16. Lu Wenzheng (2009) Characteristics of the distribution and control factors of polymetallic nodules in the western region of the CCZ (Chinese Pioneer area) In: Establishment of a geological model of polymetallic nodule deposits in the Clarion-Clipperton Fracture zone of the equatorial North Pacific. International Seabed Authority, Kingston, Jamaica, p 233–240
17. Xiang-Long J et al (1997) The characteristics of Ocean geology and deposits of polymetallic nodules in the East Pacific Ocean. China Ocean Press
18. IOM Story. www.IOM.gov.pl
19. Exon NF (1983) Manganese nodule deposits in the Central Pacific Ocean and their variation with latitude. Mar Min 4:79–107
20. Cronan DS, Hodkinson RA (1994) Element supply to surface manganese nodules along the Aitutaki-Jarvis transect, South Pacific. J Geol Soc Lond 151:391–401
21. Cronan DS (1997) Some controls on the geochemical variability of manganese nodules with particular reference to the tropical South Pacific. In: Nicholson K, Hein JR et al (eds) Manganese mineralization: geochemistry and mineralogy of terrestrial and marine deposits. Geol. Soc. Lond. Special Publication 119, p 139–151
22. Mero (1965) The mineral resources of the sea. Elsevier, p 312
23. Cronan DS (2013) The distribution, abundance, composition and resource potential of manganese nodules in the Cook Islands Exclusive Economic Zone.Tech.Rept. No 1, Cook Islands Seabed Minerals Authority, Rarotonga, Cook Islands
24. Anon (1985) Sea Technology (1985)
25. Sudhakar M (1989) Ore grade manganese nodules from the Central Indian Ocean Basin: an evaluation. Mar Min 8:201–214
26. Shama R (2010) First nodule to first mine site: development of deep-sea mineral resources from the Indian Ocean. Curr Sci 99:750–759
27. Jauhari P, Pattan JN (2000) Ferromanganese nodules from the Central Indian Ocean Basin. In: Cronan DS (ed) Handbook of marine mineral deposits. CRC Press, pp 171–195
28. Metal Mining Agency of Japan (1986–2001) Ocean resources investigations in the sea areas of CCOP/SOPAC. Annual Reports on the Joint Basic Study for the Development of Resources

29. Cronan DS, Hodkinson RA, Miller S (1991) Manganese nodules in the EEZs of Island countries in the southwestern equatorial Pacific. Mar Geol 98:425–435
30. Kinoshita and Tiffin (1993) Economic potential of nodules in Kiribati Waters. SOPAC Tech. Rept. 177 Suva, Fiji
31. Japan- Sopac Cooperative Study on Deep-Sea mineral resources in the South Pacific 1985-1994, Seafloor Atlas 1995 JICA, MMAJ, SOPAC,Suva Fiji
32. Meylan MA, Glasby GP et al (1990) Manganese crusts and nodules from the Manihiki Plateau and adjacent areas: results of HMNZs Tui Cruises. Mar Min 9:43–72
33. Bolton BR,Bogi B, Cronan DS (1992) Geochemistry and mineralogy of ferromanganese nodules from the Kiribati region of the eastern Central Pacific Basin. In: Keating BH, Bolton B (eds) Geology and offshore mineral resources of the Central Pacific Basin. Circum-Pacific Council for Energy and Mineral Resources, Earth Science Series 14, Springer-Verlag, p 247–260
34. De Carlo EH, Fraley CM (1992) Chemistry and Mineralogy of ferromanganese deposits from the equatorial Pacific Ocean. In: Keating BH, Bolton B (eds), Geology and offshore mineral resources of the Central Pacific Basin. Circum-Pacific Council for Energy and Mineral Resources, Earth Science Series 14, Springer-Verlag, p 225–246
35. McKelvey VE, Wright NA, Bowen RA (1983) Analysis of the world distribution of metal rich Sub-Sea manganese nodules. U S Geol Survey Circular, p 886
36. Clark A, Lum J et al (1995) Economic and development potential of manganese nodules within the Cook Islands EEZ. East-West Centre Report
37. Bechtel Corp (1996) Deep-Sea nodule mining prefeasability study. Bechtel Corp, San Francisco
38. Norwegian Deep Sea Mining Group (2001) Marine mineral resources. Opportunities for Norwegian Industry, p 195

Technological Developments 1980–2000

Introduction

To a considerable degree, the deep-sea mineral exploration techniques that had been developed in the 1970s were still being used in the 1980s and 90s. There was no real incentive for major developments in exploration techniques for manganese nodules as exploration for them by the US Consortia was cut back starting in the late 1970s and not long after by some of the National Consortia as well. Cobalt-rich crust exploration used many of the same techniques as were used in manganese nodule exploration. It was in exploration for the newly discovered hydrothermal minerals that major developments were needed as existing techniques for locating widely dispersed deposits such as nodules and crusts were poorly suited to finding small, localized, hydrothermal deposits on the sea floor. Most of the needed developments in exploration for hydrothermal mineral deposits would be carried out for scientific purposes by universities and government bodies as there was little interest by industry in hydrothermal deposits during the period immediately after their discovery (except in the Red Sea). An exception to this was in the case of Lockheed whose Conrad Welling, writing in the Mining Congress Journal for November 1982 [1], considered that existing deep-tow systems needed to be made free swimming so that they could cover larger areas in the search for small PMS deposits and that a deep-sea rock drill was needed with which to core them. He noted that the Bedford Institute of Oceanography in Canada had already developed a deep-sea rotary corer for basalt coring which was adaptable for PMS coring. Other requirements were in the need to better detect hydrothermal plumes in exploration for active hydrothermal systems, and, in PMS and Co-rich crust sampling, different seafloor sampling devices than were needed for nodules. Nevertheless, there were some developments in improving tools for nodule and crust exploration too, such as the Japanese's multi-frequency echo sounding (MFES) system and an updated sea floor photographic surveying technique for nodules developed by the Koreans.

D. S. Cronan, *Deep-Sea Minerals Developments in the 20th Century*,
https://doi.org/10.1007/978-3-031-52342-7_10

Seabed Mapping, Surveying and Photography

US thinking about needed deep-sea exploration developments in the 1980s included the need to explore larger areas of the sea floor than were possible using existing equipment. Accelerating work on US EEZ PMS deposits in the 1980s triggered the development of Seamark, described at the UMI in Madison, Wisconsin October 1982 and referred to in [2]. Seamark was a multibeam acoustic imaging system that was designed to be towed at 200 m above the seafloor. It was used to examine microtopography. Two additional techniques of potential use in deep-sea mineral exploration were published in 1983/84. The first, in Offshore Engineer [3] described a refined sidescan sonar system. Geoteam was claiming the first automatic mosaic presentation of sonar through a recently completed software package "computer-aided mapping sonar" (Camos). A second system which was described in Sea Technology [4] was a high-resolution sediment profiling system with a REMOTS (Remote Ecological Monitoring of the Seafloor) Camera system. This had an application in the mapping of unconsolidated seafloor mineral deposits. The camera differed from conventional underwater cameras in its ability to make vertical slices through the sediment-water interface and to photograph the deposits in profile.

French developments in the field of free-swimming submersibles were described in Sea Technology [5] a submersible discussed was named Epaulard. The objectives of this submersible were (i) to survey 30–50 km of seafloor per day with a complete photographic coverage 5 m wide, (ii), to give a precise bathymetric profile, (iii) to navigate with an accuracy better than 2% and (iv) to operate in seas up to Beaufort Force 4. Launched in 1980,it was put into service from 1982 onwards providing photographic and bathymetric data of nodule deposits for France's ocean mining consortium AFERNOD. Unlike Nautile, mentioned earlier, it was unmanned.

In Japan, an important development in manganese nodule exploration in the 1980s was the Sumitomo MFES System which operated from a surface ship. It provided information on the distribution and size of nodules on the sea floor on a real-time basis when combined with echo sounders and a sub-bottom profiler. It worked on the principle that sound waves emitted from the ship onto the sea floor undergo a backscattering effect around an area where nodules are present. The echo received back on the ship from such an area is in the form of a backscattered wave, the magnitude of which is influenced by (i) the difference between the acoustic impedance of seawater and the acoustic impedance of small nodule targets, (ii) the wavelength of the transmitting signals, (iii) the abundance of nodule targets on the seafloor and their size. The backscattered waves received on board the ship were synthetically analyzed and as a result, a rough estimate of the quantity and size of nodules on the seafloor was obtained while the ship was underway. The MFES system was used extensively in the Japanese exploration of the CCZ in the early 1980s and also later in the decade for nodule exploration in the SW Pacific under the MMAJ/SOPAC Programme.

As part of the Korean nodule program, Park et al. [6] described a new image analysis technique for the exploration of manganese nodules. Parameters such as the size distribution of nodules and their coverage of the sea floor were determined by photography. Semi-automatic procedures to extract useful features from the photographs using digital image processing techniques were developed. Thirty-five mm films were first digitized using a film scanner. The nodules on the digitized images were recognized and separated from the background based on the characteristics of the nodules. The nodule coverage and distribution of nodule diameters were then calculated from the processed images. This was a technique to aid in the processing of the massive collection of seabed photographs that had been accumulated by the Koreans by the late 1990s.

Seabed Sampling

Once deep-sea mineral deposits had been detected either visually or by remote methods, they had to be recovered to fully evaluate them. As mentioned, many of the sampling tools used in the 60s and 70s continued to be used into the 80s, but several new ones were developed. Hans Amman of Preussag AG described a hydraulic grab of one sq. meter or more in area originally developed for phosphorite sampling on the Chatham Rise off New Zealand (see Chapter "Phosphorites") and further developed and modified for PMS and Co-rich crust sampling in deeper water [7]. The grab had a TV attachment so that sampling could be observed in real-time and precisely targeted, had a high closing force for sampling hard and massive deposits, and could be emptied on the sea floor and the latter re-sampled if the initial sample was not required. The two grab jaws incorporated cutting edges with armored cutting teeth. Grab closure and opening took place via a battery-powered unit remotely controlled from the surface ship. The grab was used in the Galapagos area to sample PMS deposits in an average water depth of around 3000 m [7]. The Japanese also developed a hydraulic grab of similar design and both were used extensively in surface and near-surface PMS and Co-rich crust sampling throughout the last two decades of the twentieth century.

Surface and near-surface sampling was adequate for evaluating Co-rich crusts and the upper parts of PMS deposits but to fully investigate them, especially their thickness, coring or drilling was required. In order to sample manganese-encrusted pavements on the Blake Plateau in three dimensions, the Burke-Manheim Hard substrate Punch Corer was developed [8], which as the name implies was "punched "into the pavements by force. An advance on this was drilling and the Bedford Institute of Oceanography rock drill has already been mentioned. In the early 1980s, the US National Science Foundation commenced sponsoring research into the development of a deep-sea rock drill that would have applications in PMS evaluation. The drill was also capable of sediment coring, in-situ geological measurements, and downhole instrument implantation as well as the primary function of hard rock coring. With suitable strength oceanographic cables and winches, it was capable of operating in water depths of up to 5000 m and was designed to recover hard rock cores of up to 50 m length. The incorporation of two transponders, a video system, and a sub-bottom profiler provided accurate navigation and positioning, and the former two also provided real-time monitoring of the coring operation. Its base was outfitted with leveling capability that permitted drilling on slopes approaching 35 degrees.

Hydrothermal Plume Detection

Hydrothermal Plumes of volcanic origin are hot water discharges on the sea floor sometimes associated with PMS formation. After discharge, they rise and spread out and thus provide a large target in exploration for PMS deposits. Commonly they are many orders of magnitude larger than the vent field from which they originate. As well as being warmer than the surrounding seawater, they may contain three chemical constituents, methane, helium, and total dissolvable manganese (TDM), which commonly show increases in seawater near hydrothermal vents. The plumes could be detected better from a surface ship than from a submersible and thus a shipboard search for them was often the first phase in a PMS exploration program. However, hydrothermal plumes only emanate from active hydrothermal vents. They do not occur over extinct vents (unless there are also active vents in the vicinity) and thus exploration for them could not be used in searching for previously formed PMS deposits.

McConachy and Scott [9] described plume detection equipment that consisted of a real-time, acoustically telemetered nephelometer package deployed over a ship's side on a hydro wire. It consisted of a nephelometer for the measurement of particle concentration, a battery pack, and a pinger mounted on a frame. The upper and lower plume boundaries could be mapped by yo-yo-ing the equipment through the plume. Such equipment could also be towed behind a ship at varying depths (tow-yo-ing).

In May–June and September–October 1985 the first attempts were made to locate, map and sample hydrothermal plumes over the Southern Explorer Ridge using the above equipment. Detailed mapping over an area of known venting revealed a complex plume pattern attributable to two vent fields [9]. The plume reached a maximum height of 210 m above the bottom. Particulate matter in the plume consisted largely of iron-rich amorphous material.

References

1. Welling C (1982) The future of United States seabed mining: an industry view. Min Congress J
2. Normark WR, Morton JL, Delaney JR (1982) Open-File Report Vol. 1982 (82–200), Geologic setting of massive sulfide deposits and hydrothermal vents along the southern Juan de Fuca Ridge. US Geological Survey doi:https://doi.org/10.3133/ofr82200a

3. Ruud S (1986) Computer-aided mapping of sonar. In Oceanology, advances in underwater technology ocean science, and offshore engineering. Springer link https://link.springer.com/ https://doi.org/10.1007/978-94-009-4205-9_40
4. Germano JD (1983) High-resolution sediment profiling with REMOTS camera system. Sea Technology
5. Anon (1983) Epaulard, a deep ocean exploration vessel. Sea Technology
6. Park C-Y, Park S-H et al (1999) An image analysis technique for exploration of manganese nodules. Mar Georesour Geotechnol 17:371–386
7. Amman H (1982) Technological trends in ocean mining. Phil Trans Roy Soc Lond Ser A 307(1499):377–403
8. Anon (1983) Ferromanganese Punch Corer. Sea Technology
9. McConachy TF, Scott SD (1987) Marine mining on real time mapping. In: Proceedings of the 1987 offshore technology conference, Houston, Texas, USA, 4–7 May 1987. OTC 5719

Environment

Introduction

From not being mentioned at all by Mero [1] in the 1960s and more or less dismissed by Flipse in the 1970s [2], environmental concerns surrounding deep-sea mining attained greater importance in the 1980s and 90s. Indeed, by the end of the century, the likely environmental effects of deep-sea mining were well known, and those concerns, considerably refined, remain the same at the time of writing. The challenge was to quantify them. The general assumption in the early 1980s was that being a new industry, deep-sea mining and environmental control would proceed hand in hand, with any environmental issues being addressed by the industry before full-scale mining took place. The attitude at the time was well summarised by NACOA [3]. NACOA considered that the environmental effects of processing, transporting, and storing marine minerals could be expected to be similar to those experienced by the land mining industry. The extent to which the environment would be affected by the actual mining of deep-sea minerals would depend on the nature of the minerals themselves, the equipment and methods used to mine them, and the geographic (oceanographic) settings of the mine sites. NACOA believed that knowledge of likely environmental effects of deep-sea mining was fragmentary in the early 1980s, but with adequate care and monitoring it believed that the deep-sea mining industry had the opportunity to avoid many of the costly environmental problems that had been encountered in land mining operations, and that it should develop with simultaneous consideration of environmental, technical and legal issues.

D. S. Cronan, *Deep-Sea Minerals Developments in the 20th Century*,
https://doi.org/10.1007/978-3-031-52342-7_11

Manganese Nodules

Exploration and mining development activities for manganese nodules were at a low ebb from the mid-1980s throughout the 1990s, and thus, according to H Thiel, that was a good time to carry out environmental studies without the time pressure of impending mining (H. Thiel, Pers. Comm. 1994). Environmental studies had been carried out in limited areas of the CCZ in the late 70s and early 80s (see Part I) but these were inconclusive and pointed to the need for more detailed studies.

Thiel [4] summarised the then-perceived impacts of manganese nodule mining as follows:-

(a) the direct disturbance of bottom sediments and fauna, including at least partial destruction of the latter, by the nodule collector system and its indirect influence through re-sedimentation of the sediment plume created by the mining.

(b) the impact of this plume on pelagic organisms in the near bottom waters and on benthic fauna where the plumes re-sediment.

(c) the effects on pelagic organisms in and around the depth of the discharge of waste and sediments lifted to the surface with the mined nodules.

According to Thiel [4],most of the manganese nodules in the path of the collector would be recovered, thus removing a hard substrate habitat for organisms. Some will be missed by the collector but these are likely to be buried by the plume stirred up by the collector, thus similarly inhibiting their roles as hard substrates for organisms. The soft sediment around the nodules will be severely disturbed, compression will take place, and much of the fauna will be destroyed. Small areas may remain relatively undisturbed but these will become covered with plume sediment. Sediment surface fauna may die from starvation. Overall, the benthic community in the path of and near the collector will be tremendously altered. Thiel believed that recolonization should occur, albeit slowly, but that the re-establishment of a balanced community might take decades. Also according to Thiel [4],little was known at the time about the fate of the plume created by the collector system during the manganese nodule mining process. Large particles and aggregates would be expected to settle out of the plume and blanket the sea bed under it. However, finer material could drift away and settle out at distant locations. Trueblood and Ozturgut [5] pointed out at the 1992 UMI that sediment resuspension and redeposition during manganese nodule mining were predicted to be one of the primary impacts on benthic communities living on the abyssal sea floor. At that time the redeposition thickness that would cause significant faunal mortality was unknown. Sediment redeposition could adversely affect abyssal benthic communities through entombment, or by starvation caused by food dilution and obstructed feeding. In order to elucidate some of these issues, environmental experiments became an increasingly important part of nodule investigations during the 1980s and 1990s.

Post-DOMES Benthic Impact Investigations

As described in Part I, in 1975 the United States initiated a comprehensive research program called Deep Ocean Mining Environmental Study (DOMES). Later, beginning in 1983, NOAA in partnership with SIO and Deep Sea Ventures of the OMA consortium (see Chapter "Activities on Manganese Nodules During the Post-war Boom") commenced additional investigations on the possible effect of nodule mining on the benthic zone. According to Earney [6], deep-sea mining was thought early on to be likely to damage only the upper layer of the sea floor sediments in the path of any nodule-collecting device and destroy organisms living there. Some workers thought that if unmined strips were left on the sea floor, benthic organisms living in them would rapidly colonize the mined-out strips. Another potential problem was thought to be that plumes of suspended sediment would rise into the water column during the mining process and possibly travel considerable distances before settling. However, Lavelle et al. [7] thought that most sediments should settle within a few hundred meters of the mined areas. Heavy plume fallout could bury organisms in areas both inside and outside the actual areas mined. Earney [6] believed that sediment disturbance during deep-sea mining might not be fatal to mobile creatures, but could be fatal to sessile organisms.

There were a number of follow-up investigations carried out after DOMES in order to address some of the issues raised by the DOMES Project and to increase understanding of the possible environmental effects of deep-sea mining. These have been outlined by Morgan et al. [8].In June 1983, ECHO-1 was carried out by the Scripps Institution of Oceanography and partners in the area near DOMES site C where a small-scale mining test had been carried out in 1978 by the OMA consortium. The intent was to examine benthic recolonization by seabed sampling using a box corer, but the results were inconclusive. Later the Acute Mortality Experiment using a remote underwater manipulator added known amounts of sediment to corers positioned on the seabed with the expectation that later recovery would allow determination of the amount of sediment that would produce complete mortality of the benthic community. General conclusions were that there was little evidence of serious harm to macrofauna when buried under 1 cm of sediment, but burial under 4 cm appeared to cause the demise of 25–50% of the macrofauna within 6 days [9]. Later still, the QUAGMIRE II expedition of the SIO (April–May 1990) attempted to deploy a vehicle to sample within the mining track from which nodules were removed during the OMA pre-pilot test mining experiment in 1978. According to Morgan et al. [8], the major cruise objectives were not attained.

According to Lipton et al. [10] the first known benthic disturber system was built in the late 1970s by Sound Ocean Systems Inc. Its function was to create near seafloor sediment plumes. The device was tested off the coast of Southern California by NOAA's R/V Oceanographer. The test results were mixed because while the device seemed to have worked, NOAA was not able to track the sediment plume and little or no sediment was subsequently found in the sediment traps that had been deployed around the site. After a consultative process that started in 1982 [11], NOAA decided to pilot a second disturber system, Scripps' RUM III (Remote Underwater

Manipulator). This was one of a series of pioneering tracked deep water ROVs that were equipped with a manipulator arm and designed for a range of applications . Trials with RUM III were done in 1988, 1989, and 1990 but despite modifications RUM III "did not perform as intended" (presumably it did not make a significant plume) and it was decided to try another device. The third disturber system was formally termed DSSRS (Deep Sea Sediment Re-suspension System) [12] and was developed in conjunction with the Russian Yuzhmorgeologiya (YMG). NOAA built and supplied the disturber as well as sediment traps, current meters, and samplers, and YMG supplied vessels and crew as well as a long baseline acoustic navigation system. This was subsequently known as the Benthic Impact Experiment (BIE).

Benthic Impact Experiment

In a paper at the UMI, Trublood and Ozturgut (1992) described the Benthic Impact Experiment (BIE) [5]. The study area was in the CCZ at approximately 128 W,13 N at around 4800–4900 m depth, chosen to be near three US consortia mining claims. NOAA conducted the main portion of the BIE experiment by blanketing the experiment area with varying thicknesses of sediment using the deep-sea sediment resuspension system (DSSRS) [12]. It consisted of a 1800 kg sled, 4.8 m long, 2.4 m wide and 2.2 m high [12]. As the sled was towed across the sea floor, sediment was ingested through 45 cm wide intakes and discharged 10–20 m above the bottom. According to Sea Technology [12],one problem in conducting the experiment was in suspending the very cohesive sediment in a manner that allowed local bottom water currents to carry it out of the disturbance zone. Seventeen sediment trap and two current meter moorings were deployed to map the extent of the sediment plume created by the DSRRS. It was successfully towed 44 times through the tow zone dispersing about 1600 mt of sediment over a 1-2sq km area. Immediately following the sediment redeposition,16 multicore samples were taken to map the sediment redeposition. According to Morgan et al. [8] an important parameter that it was hoped would be attained, namely the thickness of sediment added to the seabed by the disturber, was not attained because the sediment was so widely dispersed that no significant accumulation was measurable outside of the actual area disturbed.

Also at the 1992 UMI, Dobbs and Smith [13] described the biological research consequent on the BIE investigations, carried out on macrofauna, meiofauna, sedimentary bacteria and labile protein, and nodule-associated fauna, but no relationship between faunal succession and sedimentation was obtained.

Japan Deep-Sea Impact Experiment

According to Harada and Fukushima [14], beginning in 1994, over a three-year period the MMAJ carried out the Japan Deep-Sea Impact Experiment (JET) in Japan's western manganese nodule zone in the CCZ near 9 N,146 W at a depth of approximately

5300 m,partly to assess the impact on benthic organisms of re-sedimentation. After the disturbance, sediments were recovered with a multi-corer and analyzed for water content, organic carbon, total nitrogen, calcium carbonate, and biogenic silica. The study included baseline data collection, sediment traps, current meters etc., but the Japanese also developed a quantitative photographic analysis method to try to measure the distance of sediment redeposition from the disturbance tracks. The work was divided into JET 1 which was the disturbance experiment designed to mimic the sediment disturbance that was expected to take place during manganese nodule mining and follow-up investigations to assess its impact carried out immediately afterward (JET 2), 1 year later (JET 3) and 2 years later (JET 4). During JET 1, sediment was disturbed to a depth of 7 cm. Laboratory investigations showed that in the surface layer where re-sedimentation was extensive, concentrations of calcium carbonate and organic carbon decreased and that there were changes in total nitrogen and biogenic silica too. This was hypothesized to result at least partially from the dilution of the surface layer by sediment previously buried more deeply.

Interoceanmetal (IOM) Benthic Impact Experiment

IOM carried out a benthic impact experiment in 1995. There were three surveys at the site, namely those of 1995 (immediately pre-and post-disturbance), 1997, and in 2000. Based largely on Radziejewska et al. [15], Lipton et al. [10] concluded that research at the IOM BIE site essentially (i) struggled in 1995 to find significant impact to the meiobenthic communities outside of the tracks left by the disturber, (ii) found in 1997 widespread changes (e.g. in nematode familial make-up) as a result of a phytodetritus sedimentation event which they thought was most likely to have been a natural plankton bloom event, as well as that the tracks were by then substantially " leveled off ", presumably due to action of currents or bioturbation, and (iii) found in 2000 that response to the phytodetritus sedimentation event had more or less reverted to the condition at the start of the experiment. According to Radziejewska et al. [15] the implications of this were that there is no absolute, inherent stability on the abyssal seafloor and that natural processes can induce changes of greater magnitude than induced by a BIE, and that quantitative monitoring of mining impacts will be challenged by naturally occurring variations.

Disturbance and Recolonization Experiment (DISCOL)

DISCOL was a German benthic impact experiment which commenced in the Peru Basin in 1989, close to 7 S, 88 W and near an area of German mining interest. DISCOL was intended to study the recolonization of an area artificially disturbed to replicate the sort of disturbance that might take place during manganese nodule mining [4]. The disturbance was created by a "plough-harrow" device. Twenty-four small ploughs were arranged in two rows so that 8 m wide tracks were cut into the seabed. After ploughing 78 transects crossing a 2 nautical mile diameter circular field, the sediment in about 20% of the area was turned over and much of the remainder received a cover of variable thickness through re-settling of the plume created. Bottom photographs from the DISCOL 1 expedition clearly showed the difference between the pre-impact sea floor and that post-impact. Considerable disturbance was observed. Six months later, the DISCOL 2 cruise clearly showed the ploughing tracks and diminished fauna in the area indicating a long-lasting impact. Three years later during the DISCOL 3 cruise, the picture had changed. The originally distinct tracks were more rounder and softer, probably as a result of bioturbation and currents. Recolonization of the impacted area had started. The DISCOL area was investigated again 7 years after the initial experiment [16] and continued to be surveyed with the effects of the original disturbance still being visible into the twenty-first century.

Indian Deep-Sea Environmental Experiment (INDEX)

According to Sharma [17] the Indian Deep-Sea Environmental Experiment was a benthic disturbance experiment conducted in the Central Indian Ocean Basin to study the likely environmental impact of manganese nodule mining there and to assess restoration and recolonization processes. The objectives of the program were to (a) establish marine environmental baseline conditions in the Indian manganese nodule mining area, (b) assess the likely impact of nodule mining on the marine ecosystem by simulating a benthic disturbance, (c) to understand the processes of restoration and recolonization of the benthic environment, and (d) to provide scientific input to the design of a nodule mining system. A benthic disturbance and impact experiment was carried out during July and August 1997. A 3000 m long and 200 m wide strip was selected for the experiment using the type of deep-sea sediment resuspension system (DSSRS) used in the BIE. Analysis of the seabed after the experiment showed disturbance features such as sediment piles and re-sedimented areas in and around the disturbance site where the benthic activity was very low. No effects of the plume were observed from CTD and rosette profiles at 20 m above the seafloor 5 days after the completion of the disturbance suggesting that the bulk of the disturbance plume settled quickly or moved away. A reduction in macrofaunal abundance as well as in meiofauna and bacteria indicated partial destruction of benthic life immediately after the disturbance. However, increases in some biochemical components (nutrients) in the sediments post-disturbance were considered to point to a likely increase in faunal density in the future. Overall, the results of the experiment showed vertical mixing of sediment, lateral migration of the sediment plume, changes in the physical, chemical and biochemical properties of the sediments and an overall reduction in benthic biomass.

Synthesis

Morgan et al. [8] in their synthesis of the environmental effects of deep-sea mining, based on the experiments outlined above, concluded that all studies up to that time supported the initial concerns about the environmental impacts of nodule mining that were identified in the original DOMES research, principally disturbance and at least partial destruction of benthic communities within a nodule mining area, and impacts on benthic communities at some distance away due to the re-deposition of suspended sediments. However, they pointed out that none of the studies carried out during the 80s and 90s had been able to establish firm quantitative relationships between burial depth by resuspended sediment and impacts on fauna. Furthermore, none of the post-DOMES work significantly illuminated the effects of surface discharge of waste from nodule mining ships such as possible impacts on the near-surface biota of the discharge of "foreign materials" like nodule fragments and associated sediments into the water column. These and other environmental aspects of nodule mining of concern only became subjects of research in the twenty-first century and are still underway at the time of writing.

Polymetallic Sulphides

Until the late twentieth century,the only PMS deposits that had been considered for possible mining were those in the Red Sea. These had been discovered in the 1960s,evaluated in some detail in the 1970s and have been described in Chapter "Hydrothermal Deposits: Discovery and Preliminary Economic Evaluation".Unlike in the case of manganese nodules where environmental considerations were given little or no thought in the early days, they were prominent in Red Sea minerals activities almost from the start. Environmental studies ran on after the preliminary exploration phase was over as longer term monitoring than could be accomplished during the exploration phase was needed. This was the MESEDA (Metalliferous Sediment Atlantis II Deep) campaign.

The MESEDA campaign has been reviewed by Thiel [18]. It covered several cruises focussing on the Atlantis II Deep. Environmental impact studies both on the sea floor and in the water column overlying the Atlantis II Deep were carried out. These included, among other things, studies on circulation, currents, upwelling, turbidity, temperature, salinity, dissolved gasses and nutrients, phytoplankton, primary productivity, benthos abundance, heavy metals, mineralogy, chemistry and toxicology of tailings, plume development and the effects of tailings on the marine environment.

According to Thiel [18], in order to avoid impact on the upper waters, a tailings discharge depth of 400 m was adopted. The tailings discharge was monitored in the water column using a 30 kHz echosounder among other things. Tailings containing 225 tonnes of particulate material were discharged. The plume was traced down to 1100 m. After 10 days the plume had a horizontal expansion of 5000 m. It was estimated that over the expected lifetime of an Atlantis II Deep mine, approximately 50 million tonnes of waste would be discharged over a period of 15–20 years. The tailings were found to be toxic. This was thought likely to affect the benthos, together with the blanketing effect of settling particles. Thiel [18] concluded that while the MESEDA programme taught us much about the oceanography of the Red Sea, it was insufficient for a definite evaluation of mining impacts there. After a long period of quiescence, environmental studies relating to metalliferous mud mining in the Red Sea resumed in the second decade of the twenty-first century (see Epilogue).

Serious attention was not given to mining open ocean PMS deposits until the twenty-first century and thus there were no environmental experiments on them in the twentieth century. An up-to-date review of actual and needed environmental studies around PMS deposits, in general, has been given by Van Dover et al. [19] together with references therein. At the time of writing, no mining of deep-sea PMS deposits has taken place anywhere.

References

1. Mero JL (1965) The mineral resources of the sea. Elsevier, p 312
2. Flipse JE (1983) Deep-ocean mining economics. OTC Houston
3. NACOA (1983) Marine minerals: an alternative mineral supply. National Advisory Committee on Oceans and Atmosphere, July 1983, Washington, DC
4. Thiel H (1994) Mineral mining in the deep sea: Environmental consequences and precautionary research. II International Conference on Oceanography, Lisbon,14–19 November 1994, IOC/OCEANS/WD/41
5. Trueblood DW, Ozturgut E (1992) The Benthic Impact Experiment (BIE): a study of environmental impacts of manganese nodule mining on the abyssal sea floor. Underwater Mining Institute
6. Earney FCF (1990) Marine mineral resources. Routledge, p 385
7. Lavelle JW, Ozturgut SA et al (1981) Dispersal and re-sedimentation of the benthic plume from deep-sea mining operations: a model with calibration. Mar Min 3:60–90
8. Morgan CL, Odunton NA, Jones AT (1999) Synthesis of environmental impacts of deep seabed mining. Mar Georesour Geotechnol 17:307–356
9. Smith CR, Bennett BA, Brumsickle SJ (1988) Assessment of benthic faunal sensitivity to rapid sediment burial at DOMES Site C-1.Final Report to NOAA/OME Seattle. University of Washington
10. Lipton I, Nimmo, M, Parianos J (2016) TOML Clarion-Clipperton zone project, Pacific Ocean. NI43–101 Report. Brisbane Australia, AMC Consultants Pty Ltd. Available at www.sedar.com
11. Ozturgut E, Trueblood DD, Lawless J (1997) An overview of the United States's Benthic impact experiment. In: Proceedings of international symposium on environmental studies for deep-sea mining. Metal Mining Agency of Japan, Tokyo, Japan, 20–21 November 1997
12. Brockett T, Richards C (1994) Deepsea mining simulator for environmental impact studies. Sea Technology, pp 77–82
13. Dobbs FC, Smith CR (1992) Evaluating benthic community disturbance and succession following simulated manganese nodule mining in the equatorial Pacific. Underwater Mining Institute

14. Harada K, Fukushima T (1997) Results of seabed disturbance experiment and bottom sediment investigation. In: Proceedings of international symposium on environmental studies for deep-sea mining. Metal Mining Agency of Japan, Tokyo, Japan, 20–21 November 1997
15. Radziejewska T, Rokicka-Praxmajer J, Stoyanova V (2001). IOM BIE revisited: meiobenthos at the IOM BIE site 5 years after the experimental disturbance. In: International society of offshore and polar engineers 4th ocean mining symposium, Szczecin, Poland
16. Schriever GA, Ahnert A et al (1997) Results of the large-scale environmental impact study DISCOL during eight years of investigation. Proc 7th Int. offshore and polar engineers conference, p 438–444
17. Sharma R (2001) Indian deep-sea environmental experiment (INDEX): an appraisal. Deep Sea Res II 48:3295–3307
18. Thiel (1991) From MESDA to DISCOL; a new approach to deep-sea mining risk assessments. Mar Min 10:369–386
19. Van Dover C, Colaco A et al (2020) Research is needed to inform environmental management of hydrothermally inactive and extinct polymetallic sulfide (PMS) deposits. Mar Policy 121:104183

Economic and Legal Developments, 1983–2000

Introduction

The difficulties faced by potential deep-sea miners as a result of the declining prices for the main nodule metals in the late 1970s, the generally unfavourable (for them) outcome of the Law of the Sea Convention, and increasing environmental concerns associated with marine mining, engendered a change in the perceived worth of deep-sea minerals. Long gone were the over-optimistic assessments of John Mero and his cohorts. Instead, a new realism took over exemplified by the work of James Broadus of the Woods Hole Oceanographic Institution. According to Broadus [1], potential seabed mineral resources would have to be viewed in relation to both conventional and more speculative alternative mineral resources on land, or wait to be developed when little or nothing of those resources remained. Broadus considered that the best approach to the identification of seabed minerals as future resources was a "long-run supply function" for each mineral. These were predicated on the amounts of each mineral that could be obtained economically at different levels of incremental unit cost. Broadus believed that for most seabed minerals, continuing onshore production from successively costlier deposits would be expected until they became so expensive that the least cost seabed equivalents could be mined in competition with them. Once that threshold had been reached, the division of output between onshore and offshore mineral resources would depend on their respective available quantities for each increment of elevated cost. The difficulty, however, in applying this economic model over the subsequent decade and a half was that apart from minor ups and downs during this period the prices of the most important metals did not increase, and indeed actually decreased (Fig. 1).

D. S. Cronan, *Deep-Sea Minerals Developments in the 20th Century*,
https://doi.org/10.1007/978-3-031-52342-7_12

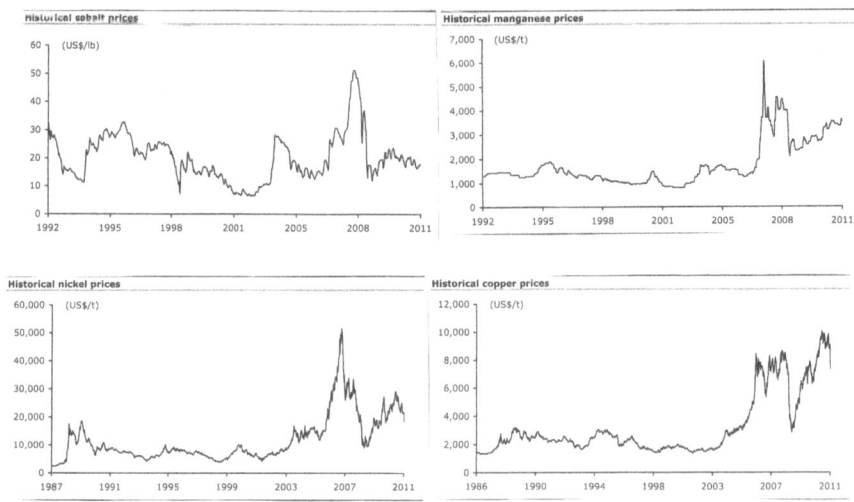

Fig. 1 Metal prices at the end of the twentieth century and into the twenty-first

Hydrothermal Minerals

Unlike both nodules and crusts which are widely distributed compositionally rather uniform deposits on the sea floor, hydrothermal deposits are localized deposits of very variable composition. This fact considerably influenced initial attempts to carry out an economic evaluation of them and made a realistic estimate of their value other than in the Red Sea (Chapter "Hydrothermal Deposits: Discovery and Preliminary Economic Evaluation") very difficult. One of the first was by Bischoff et al. [2]. They pointed out that much of the then interest in PMS was based on their content of potentially valuable metals. For example, the Cu-rich Galapagos Rift deposit was thought to be an example of Cyprus type massive sulphides. They concluded that major unanswered questions concerning the exploitability of PMS deposits at that time lay with uncertainties regarding their size, degree of consolidation and distribution. These inhibited a realistic economic evaluation of them.

In the mid-1980s, an attempt was made by Broadus [3] to put the economic potential of hydrothermal polymetallic sulphides (PMS) into perspective. However, like Bischoff et al. [2] he concluded that too little was known about them to permit confident estimates of their economic worth. Nevertheless, based on consumption projections he thought that it could be possible long-term scarcity of the metals contained in PMS that might make them of economic interest in the distant future. He tended to agree with Rona [4] that most of the PMS deposits found in the Pacific up to that time fell into the category of "mineralized showings" rather than of potential resources. He mentioned the general lack of interest in the deposits at that time by most of United States industry, other than those companies that stood to gain most from exploration activity by developing and selling technology.

Another attempt to review the economic potential of PMS deposits was by Glasby [5] who opined that Private Industry had no interest in exploiting actively forming hydrothermal PMS deposits at that time but thought that it might be interested in older extinct deposits. He also drew attention to what he considered to be some of the exaggerated claims of earlier workers. He pointed out that Malahoff [6] had described the largest deep-sea hydrothermal mineral deposit that had been found up to that time, which was in the Galapagos Rift in 2850 m water depth. The deposit was said to be over 1000 m long, 35 m high and up to 200 m wide with a density of 5.8 and contained up to 10% Cu, 35% Fe and an average of 1% Zn, corresponding to a deposit of about 20–25 million tonnes. However, Glasby pointed out that this did not accord with the more detailed description of the deposits given by Malahoff et al. [7]. Instances such as this confirmed Broadus's view that in the mid -1980s there were simply not enough reliable data available on PMS deposits to attempt a meaningful economic evaluation of them. Even later in the 1980s and in the 1990s when more data were available, estimates of global PMS deposit worth were not really possible because of the heterogenous nature of the deposits and their irregular distribution on the sea floor.

One feature of PMS deposits that was thought to be able to enhance their value was their precious metal content, particularly gold [8]. According to Hannington et al. [9] in a paper given at the UMI, by the mid-1990s more than 100 occurrences of seafloor hydrothermal mineralization had been documented but large accumulations of PMS deposits were still only known at fewer than 20 sites. Nevertheless, chemical analyses of samples from these deposits indicated that they contained significant gold and silver concentrations, comparable to those in massive sulphides on land. Gold and silver concentrations were found to be locally high in both back-arc and mid-ocean ridge settings. For example, estimates of three–five million tonnes of PMS at Explorer Ridge in the Canadian/US EEZ, by Scott et al. [10], with average concentrations of 0.8 ppm Au and 106 ppm Ag, suggested possibly 70,000–120,000 oz. of contained Au and 9–15 million oz. of Ag in the deposit. Although base metals remained the main resource of interest in PMS deposits, precious metals could have contributed as much as 20% of their value.

Manganese Nodules

In a 1984 article entitled "Commercial exploitation of polymetallic nodules-when will it start" Jan Magne Markussen of the Fridtjof Nansen Foundation in Norway addressed this question and provided a useful summary of the situation in regard to possible manganese nodule mining in the immediate post-Law of the Sea era [11]. Briefly reviewing the activities of the US commercial consortia in the 1970s, he summarised estimates that showed that CCZ nodules of interest to the consortia contained amounts of Ni, Cu, Co and Mn that were equal to approximately 25, 5, 180, and 20% respectively of the then land-based reserves of those metals. He also noted that estimates made by the French indicated the CCZ had nodules sufficient for up to ten first-generation mine sites that would each have a production capacity of three million tonnes of dry nodules per year over a period of 20–25 years. However, he also noted that after the US industrial consortia had completed their initial R & D programmes in the late 1970s, their level of activity reduced. Indeed, he pointed out that it was only the OMA Group that continued significant development and testing into the 1980s and that it was able to do so because an Italian company had invested between 20 and 25 million dollars in the OMA Group in 1980 to obtain an ownership share of 25%. Markussen then went on to discuss the causes of the reduced level of activity of the US Consortia. These were considered to be due to both economic and legal factors. The then low prices of the metals of interest impeded development of a nodule mining industry. Also, he noted that discontent with parts of the system of rules for mining operations laid down by the 1982 Law of the Sea Convention, together with a "general mistrust of a management system based on the United Nations" had been mentioned by the consortia as factors hampering development. Markussen concluded that commercial exploitation of nodules was characterised by uncertainty as to both when it would start and, once started, how it would proceed.

In the late 1970s and throughout the 1980s, one factor of particular concern in the United States was the need to ensure an adequate supply of strategic minerals. Because of this NACOA, the National Committee on Oceans and Atmospheres (a public body that advised the government on matters within its remit) examined the domestic mineral supply situation [12]. Metals that it considered to be "strategic" included Cr, Co, Ni, Mn, Pt, Ti, W and V. Under the US Strategic and Critical Minerals Stockpiling Act of 1979, strategic minerals were defined as "those minerals which would be needed to supply the military, industrial and essential civilian needs of the United States during a national emergency and are not found or produced in the United States in quantities to meet such needs". This was the time of the cold war and the then-recent disruption in Zaire and its effect on the price of cobalt focused attention on this problem. The circumstances associated with the then-current mineral supplies, and projections of likely future demands, were reviewed to see if the development of marine minerals could reduce dependence on mineral imports. Consequent on this review the Committee believed that the development of a national marine minerals industry should be a national goal and concluded that the absence of favourable economic and legal conditions for marine mining at the time should not forestall the US from taking measures to recover seabed minerals as they might be needed in the future.

In June 1980, President Carter signed the Deep Seabed Hard Minerals Act which authorised the licensing of US firms to mine beneath international waters until such a time as international covenants were approved. NOAA managed the Act and received reports from the licensees. Following the decision of the USA to vote against the Law of the Sea Convention, the certification of four US applicants for the first deep seabed mining exploration licences under the Act was announced by NOAA in 1984. All the licence areas were in the CCZ. NOAA called this certification an important step in recognising the site security for US companies to pursue mineral development in international waters. Their locations were published in issues of the Federal Register in late 1984. Kennecott received an additional licence for one site from the UK and OMI received additional licences for two sites from Germany, under deep seabed mining legislation of those countries which were "reciprocating states" (see below) harmonised with the US Deep Seabed Hard Minerals Act.

According to Lipton et al. [13], in the 1980s efforts to manage deep-sea mining (and specifically development of the CCZ nodules) split. As outlined in Chapter "The Post-second World War Deep-Sea Minerals Scene", the United Nations established a Preparatory Commission (Prep Com) to look at developing regulations to implement the 1982 Law of the Sea Convention and to encourage Pioneer Investors into the Area. Many of the groups that had worked in the Area in the 1970s wanted to register and protect past work in the hope that a Law of the Sea Convention favourable to them was imminent. These were designated as Pioneer Investors. There was also an attempt to establish a Reciprocating States Regime (RSR), which initially involved the USA, Japan, France, West Germany, and the United Kingdom. The RSR was designed to protect the mutual rights of its members principally in regard to overlapping claims, and was planned either as a bridging system until UNCLOS was concluded, or could be an alternative system altogether. Suspicion existed between leading proponents of UNCLOS (such as the USSR) and proponents of the RSR. After several attempts to progress discussions, the USSR and India registered first as Pioneer Investors in early 1984. France and Japan followed suit in late 1984. Issues with the applications, in part relating to overlapping claim areas (see below), meant that most of these applications were resubmitted later along with others. The Pioneers received special terms including application fees at cost of administration (up to USD 250,000) and the right to apply for 150,000 sq. km in the first instance.

Also according to Lipton et al. [13] the RSR worked by mutual cooperation. States established domestic deep-sea mining legislation of a broadly similar nature and agreed that mineral rights granted under their respective domestic legislation would not overlap. Overlapping claims had come about because of the way in which early nodule exploration in the CCZ had been done, individual groups working independently to find the best areas. Numerous rounds of negotiation were required between the parties concerned to eliminate overlapping claims, including the exchange of data. As mentioned, an "Interim Agreement" was signed between the United States, France, West Germany, Japan and the United Kingdom in 1982 and this was replaced in 1984 by an expanded "Provisional Understanding" that also included Belgium and Italy [14]. Subsequently, consultations between the US, Belgium, Canada, Germany, Italy, Japan, the Netherlands, the UK and the USSR produced a mechanism for the resolution of remaining mine site overlaps. Naturally, those organisations that entered the nodule exploration business later such as Korea, China and the IOM Consortium were not party to these agreements and had to find areas to mine that had not been claimed earlier. By 1993 there were 18 "operating areas" [13]. Most of these areas are now encompassed within ISA contracts.

Looking at nodules as more than just a source of metals, William Siapno, who had earlier worked on the Deep Sea Ventures nodule program, outlined at the UMI another aspect of the then thinking on the benefits of manganese nodule mining. This was their possible use as "gas scrubbers" [15]. This idea was not new but increasing air pollution related problems in the United State in the 1970s and 80s brought it back into focus. Nodules were thought to possess sufficiently high surface areas and transition metal contents to be of possible use in this regard on a larger scale than had been considered before. Again, but on a different note, Siapno [16] gave a paper at the UMI on the potential use of nodules as a battery component. Siapno proposed the use of batteries containing manganese derived from nodules to drive electric cars. This was most perceptive in view of developments in the twenty-first century relating to the role of nodules as a source of battery elements.

In 1986 the UN published a report on methodologies for assessing the impact of deep-sea minerals on the World Economy, based largely on the work of the US manganese nodule consortia in the 1970s and 1980s [17]. In this report was a review of existing US quantitative studies dealing with seabed minerals, some of which have been dealt with in Part I of this book. One additional point made was just how uncertain were many of the assumptions made about deep-sea mining at that time. For example, the technology to be involved was new and largely untested under expected operating conditions. Also, the number of mining operations that could be sustained was unclear and would depend, among other things, on access to deposits, technology and finance. Other uncertainties were related to the demand for and prices of the principal nodule metals of economic interest at that time, namely Ni, Cu and Co. A crucial variable was the likely size of ocean supply relative to the existing size of the market for the metals concerned. For example, the report reiterated that profitable recovery of Ni would have produced more Co than could be absorbed by the then market which could have negatively affected the price of Co. When the Law of the Sea Convention came into force, measures, including production quotas, were expected to be taken to protect the economies of land-based producers of the minerals concerned (see Part I). The UN study concluded that deep-sea mining would be a risky venture whose investors would require a risk premium as well as reasonable long-term stability in the institutional arrangements. This was not thought to be a problem for developed country participants but could present difficulties for participants from developing countries. All of this demonstrated considerable uncertainty regarding the development of deep-sea mining in the mid-1980s, even assuming that metal prices would be conducive to it, which they turned out not to be.

In 1990, John Padan, then with the US Minerals Management Service and formerly deep seabed mining programme manager at NOAA, wrote an article on work on nodules carried out by the US-based consortia, and others, during the previous 20 years, largely based on the experiences of NOAA licensees as revealed in their applications for licences, and in their annual reports to NOAA covering the first 4 years of their licence terms [14]. He also gave maps detailing the licence areas and those areas reserved for the ISA, which were to be the focus of manganese nodule activities early in the twenty-first century. A principal objective of all the consortia holding licences issued by NOAA had been to learn whether or not their licence sites contained enough recoverable resources to satisfy their need for a nodule production of three million tonnes annually for 20 years. By 1987 such had been established by the Kennecott Consortium (KCON) but the remaining NOAA-licenced consortia reported that they still needed additional data on nodule abundance, variability and location of obstructions in their areas, to satisfy their own standards for being certain of the amount of recoverable ore. However, Earney [18] reported that largely due to the Consortia's increasing pessimism regarding deep-sea mining, as the 1980s progressed they either modified their licences or gradually withdrew such that there was very little activity of any sort on manganese nodules by those consortia by 1990.

One development that was at variance with the prevailing view at the time was a French study [19] that purported to show that a French manganese nodule mining operation could be profitable in the late 1980s. It would have involved a 1.5 million tonnes per year recovery rate, would have involved processing in France and would have yielded an IRR of 12% on the basis of Ni at $3.60 per pound, Cu at 0.95$ per pound and Co at $6.80 per pound. Such a return was to be achieved by "improved metallurgical processing" and the use of nuclear power generation. Such optimistic predictions were uncommon at the time, and, indeed, a French nodule mining operation did not take place then, or subsequently.

Throughout the 1980s and into the twenty-first century, the East-West Centre, on the Manoa campus of the University of Hawaii, played a prominent role in analysing deep-sea minerals developments, especially in the Pacific. In 1990, Charles Johnson, research associate at the Centre, produced a summary of Centre views on deep-sea minerals and mining for testimony before US Government Committees. Johnson addressed some the issues current at the time, including:

(i) Why it was believed that US industry was not interested in deep-sea mining,
(ii) The prospects for deep-sea mining in 10–15 years,
(iii) Technological possibilities in deep-sea mining and
(iv) Why the US Government's attitude would not result in US leadership in any future deep-sea mining.

Johnson believed that there were two main reasons why US Industry was not interested in deep-sea mining, the risks were too high for the modest expected return on investment and the planning horizon was too long. It seemed to him that deep seabed mining by US entities would only occur in the then foreseeable future if there were to be substantial government R&D funding available, which there was not. Johnson noted that new land-based mineral deposits were continually being found, more than offsetting the then rates of metals depletion. Also, with the winding down and subsequent ending of the Cold War in the late 1980s and early 1990s, there was less of a need for stockpiling "strategic" minerals. Nevertheless, Johnson was optimistic that deep seabed mining would take place early in the twenty-first century. The reason for this was mainly that when the technological advances made in the 1980s, and those expected in the 1990s, were factored in, he believed that there would be a substantial improvement in the economics of deep-sea mining. He believed that the technological advances being made in robotics and systems control would eventually allow the development of multiple independent mining units for the sea floor. Mining units were thought to be only a small share of total mining costs and thus major increases in production might be possible for only a modest increase in investment. Johnson reiterated the view that most of the capital and operating costs for nodule mining projects were in the processing, and in his view, these would be reduced in the future with technological improvements. However, he believed that previously suggested savings in transport costs to Pacific Island-based processing plants near the mining sites, rather than transport to continental sites further away, would be more than offset by savings in energy, water and waste disposal costs in the latter. Finally, echoing the concerns of Moore [20] outlined in the Introduction to Part II of this book, Johnson pointed out that the main problem appeared to be in the US Governments' failure to support innovative technology development to reduce technical risk and shorten the time horizon for commercial development of deep-sea minerals. Johnson's view was that the US which had been the technology leader in deep-sea mining developments in the 1970s had fallen behind other countries such as Germany, France and Japan in the 1980s, and was not expected to catch up.

The Ni deposit at Voisey Bay in Canada is not a marine deposit, but its discovery in the early 1990s had a considerable influence on the development of a deep-sea mining industry for manganese nodules. As if things were not already bad enough for the potential manganese nodule mining industry in the late twentieth century, the discovery of a large accessible Ni, Cu, Co deposit on land added to the difficulties. The deposit was not actually mined until 2005, but from the 1990s onwards its presence was a damper on the development of a marine manganese nodule mining industry. The deposit was estimated to contain about 140 million tonnes of ore, producing a Ni, Cu, Co concentrate (the same metals as in nodules but in different relative proportions). The discovery of the Voisey Bay deposit was definitely a major obstacle on the road to deep-sea mining, at least for Ni, Co and Cu.

Cobalt-Rich Crusts

Many of the issues of importance in regard to nodules also applied to Co-rich crusts. However, there were important differences. Perhaps the greatest is that crusts are firmly attached to the seafloor whereas nodules are loose and thus crusts would have been much more difficult to recover economically than nodules. This dominated economic thinking on crusts during the last part of the twentieth century. Halbach and Manheim [21] discussed some of the economic implications of their work on Pacific seamount crusts in the early 1980s and estimated that the value of the metals in the crusts in the areas investigated significantly exceeded those in prime deep water nodule areas. Furthermore, many of the encrusted seamounts were sufficiently large to meet the tonnage requirements indicated to serve as a minimum for first-generation nodule mining sites.

Unlike in the case of the possible mining of manganese nodules which had been the subject of several economic studies by the mid-1980s, the economic evaluation of possible crust mining was in its infancy at that time. A joint US Federal/State task force was formed in the early 1980s to study the economic potential and environmental impact of the recovery of crusts in the 200-mile EEZ surrounding the Hawaiian Archipelago. The initial assignments of the task force were to delineate the main areas of interest for exploration and to define those environmental concerns which needed to be taken into consideration in an Environmental Impact Statement. As mentioned in Chapter "Expanding Cobalt-Rich Crust Activities in the Pacific Ocean", an assessment of the economic potential of crusts in the EEZ not only of Hawaii but additionally of Johnson and Palmyra Islands, and elsewhere, was given by Clark et al. [22]. Prior estimates indicated that the amounts of metals available were quite large relative to their total US consumption and be able to supply US requirements for Co and Mn for more than 75 and 10–20 years respectively at the then rates of consumption. Crust mining was also thought to be able to provide a higher value recovery than nodules, up to $11 per square meter for the crusts but only $ 1.27 per square meter for the nodules. It was thought that the main constraint on the size of crust mining operations would be the size of the Co-market and the price of Co. The supply of Co from crusts would depress Co prices and that one crust mining operation would preclude any others for a significant amount of time. Manheim [23] further considered the economic value of crusts and asked the question "how would crust production compare with US domestic metal demand". He suggested that even one crust operation of one million tonnes could have met a significant part of Co demand, about one-quarter of Mn demand and small fractions of the demand for other metals such as Ni, Cu, Pt, V and Mo.

Updating earlier crust work John Wiltshire, Director of the Hawaii Undersea Research Lab (HURL), considered more recent work on crusts from the Hawaiian and Johnson Island EEZs [24]. Resource assessment was undertaken on six research cruises off the Hawaiian Islands and one off Johnson Island. Analysis of the several hundred crust samples brought back from these cruises showed over 1% Co in several cases. In addition to the main metals of economic interest in the crusts (Co, Ni and Mn) Wiltshire drew attention to the Pt, Pd and Au in them. Sixty-nine Hawaiian and Johnson Islands crust samples contained Pd in amounts of 10–20 parts per billion (ppb). Gold reached levels of 500 ppb but generally fell in the 100–200 ppb range. Platinum was the most abundant of the three metals with values as high as 2 parts per million (ppm). Platinum appeared to be most favourably concentrated in crusts south of Johnson Island, and in these Wiltshire believed its recovery could have enhanced their value. Indeed, a preliminary economic analysis reported by Wiltshire indicated that Pt concentrations in the range of 1–2 ppm would increase the value of a crust resource by 20–40%. Nevertheless, it was unlikely that crusts would ever have been recovered just for their Pt. It would have had to have been a by-product.

An updated economic evaluation of possible Co rich crust mining was given by Wiltshire [25]. He pointed out that compared to many other metals markets, the Co-market was more volatile, less predictable and, at that time, highly lucrative for Co-producers. There were a number of elements of the Co-market which caused this, including its relative smallness, the wide variety of products which used Co, and the relatively small number of producers. This situation was complicated by a significant growth in Co demand then ongoing and projected to accelerate over the following several years. A significant number of new terrestrial Co deposits were actually Ni-Co deposits from which Co would be recovered as a by-product, but although the world Ni market was expanding it was not expanding as fast as the Co market. This limited the number of deposits that could be brought into production at any one time. It was in this context of growing demand for Co that the potential of a market share for a crust mining venture was considered. Probably a successful crust mining venture would have needed to sell between 2 and 5 thousand tonnes of Co per year.

Also according to Wiltshire [25], in 1996, the world Co market was actually a series of small independent markets. None of these markets was declining at the time; all were either growing or maintaining their then levels. Most of the Co usage was high tech and many of the applications had prospects for considerable growth. Overall, long-term demand for Co appeared to be increasing at 3–4% a year. The then world demand for Co and the available supply were in equilibrium. This balance was maintained by several factors. One of these was sales from the US Co stockpile which had been built up during the Cold War in case the US was cut off from Co supply. As this was no longer needed in the 1990s it was being sold off and was an important component of the then world Co market. Another factor maintaining the then supply/demand balance of Co was new sources of Co either becoming, or soon to become, available, such as the deposit at Voisey Bay (see above). However, these terrestrial deposits were, as mentioned, primarily Ni deposits. Crusts had an advantage that they were primarily a Co deposit. Since the rate of expansion of the Ni market was unlikely to be large enough to supply enough by-product Co to satisfy the expected increasing demand for it, it is hardly surprising that crusts were given serious consideration as an alternate Co source as the twentieth century drew to its close.

Law of the Sea

In the aftermath of the 1982 Law of the Sea Convention, there was much discussion both in the USA and elsewhere on its implications for seabed mining.

According to Padan [14], by the late 1980s, in the light of reduced profit expectations, the US-based deep-sea mining consortia wanted to renegotiate their obligations under the Law of the Sea Convention if and when the USA might accede to it. The less developed countries who were the main beneficiaries under the original deep-sea mining provisions of the Convention were apprehensive, resulting in protracted discussions on a way forward. Although many States had signed it, the 60 ratifications necessary for the Convention to enter into force were coming from developing States. As the number of required ratifications approached, with only Iceland as the sole developed ratifying State, leaving it and the 59 developing States-Parties to among other issues, shoulder alone the considerable financial burden for, e.g., the ISA, the pressure to resolve the situation increased. During this period there was little activity in nodule mining development. According to Lipton et al. [13], the UN Secretary-General considered the idea of modifying Part XI of the Law of the Sea Convention, that part dealing with deep-sea minerals and mining. Discussions began in July 1990 and advanced through progressively widening consultation with the States. In 1994, with the Convention coming into force in November of that year, proposed revisions to Part XI and the Annexes had progressed to the point of widespread acceptance. The resulting Implementing Agreement of July 28th, 1994 (the IA) mandated that key articles of the original Convention, including (i) those of greatest concern to future deep-sea miners on limitation of seabed production (the quotas) and mandatory technology transfer, would not be applied, (ii) the United States if it acceded to the Convention, would be guaranteed a seat on the Council of the pending International Seabed Authority, and (iii), voting would be done in groups, with each group able to block decisions on substantive matters. Nevertheless, the IA despite its being drafted largely to accommodate the United States, did not lead to the United States acceding to the Convention, although it did sign, but not ratify, the IA, which came into force in July 1996.

With the entry into force of the Law of the Sea Convention in 1994, the International Seabed Authority (ISA) formally came into existence. It started working as an autonomous international organization, domiciled in Kingston, Jamaica, 2 years later in June 1996. All parties to the Law of the Sea Convention are members of the ISA. Odunton [26] defined its three principal organs and two subsidiary organs, namely (i) The Assembly is the supreme organ of the Authority, containing all its members (ii) The Council is the executive organ of the Authority, it has 36 members in five geopolitically arranged chambers that are periodically elected by the Assembly, (iii) The Secretariat is the administrative organ of the Authority and headed by the Secretary-General. Amongst other things, the ISA maintains a legal advisory function and a central data repository. The Legal and Technical Commission (LTC) makes recommendations to the Council on most matters, including regulations, applications, and environmental protection measures. It is mandated to consider and recommend approval (or otherwise) of work programs. Its minimum of 15 members are elected by the Council from nominated candidates every 5 years. The ISA has four main functions, (i) to administer the mineral resources of the seabed in the Area, (ii) to enact rules, regulations and procedures relating to these resources, (iii) to promote and encourage marine scientific research and development in the Area, and (iv) to protect and conserve the natural resources of the Area and prevent significant damage to the environment.

The revised seabed mining regime that resulted from the 1994 IA offered mining entities some of the legal certainties needed for large-scale long-term investments. It attempted to provide a more favourable legal regime for deep-sea mining, at least by the private sector, than that in the original Convention. However, difficulties such as the discovery of the Voisey Bay Ni-Co deposit and continuing depressed metal prices counteracted its beneficial provisions and no new deep-sea nodule mining development took place during the remainder of the twentieth century. As pointed out by Crockett et al. [27] it had been the "softness" in demand for metals that had been the main problem in inhibiting the development of deep-sea mining, even more so than the provisions of the Convention. At the end of the twentieth century, the outlook for deep-sea mining remained unclear. It was thought to be likely to start sometime in the twenty-first century, although there was no agreement as to when. What was clear, however, was that with so much momentum towards deep-sea mining having been built up, the possibility of it just being abandoned was remote.

References

1. Broadus JM (1987) Seabed materials. Science 235:853–860
2. Bischoff JL, Rosenbauer RJ et al (1983) Seafloor massive sulphide deposits from 21 N, East Pacific Rise, Juan de Fuca Ridge, and Galapagos Rift: bulk chemical composition and economic implications. Econ Geol 78:1711–1720
3. Broadus JM (1984) Economic significance of polymetallic sulphides. In Offshore mineral resources, Proceedings of the 2nd international seminar, 19–23 March 1984, Brest France, p 559–576
4. Rona PA (1983) Potential energy and mineral resources at submerged plate boundaries. Nat Res Forum 7:329–338
5. Glasby GP (1986) Marine minerals in the Pacific. Oceanogr Mar Biol Ann Rev 24:11–64
6. Malahoff A (1982) A comparison of the massive polymetallic sulphides of the Galapagos Rift with some continental deposits. Mar Tech Soc J 16(930):39–45
7. Malahoff A, Embly RA et al (1983) The geological setting and chemistry of hydrothermal sulfides and associated deposits from the Galapagos Rift at 86 W. Mar Min 4:123–137
8. Hannington MD, Peter JM, Scott SD (1986) Gold in seafloor polymetallic sulphide deposits. Econ Geol 81:1867–1883
9. Hannington MD, Franklin JM et al (1994) Base and precious metal resources in seafloor polymetallic sulphides. Underwater Mining Institute
10. Scott SD, Chase RL et al (1990) Sulphide deposits, tectonics and petrogenesis of southern Explorer Ridge, Northeast Pacific Ocean. In: Malpas J, Moores EM et al (eds) Ophiolite-Oceanic Crustal Analogues. Geological Survey Dept, Nicosia, pp 719–733
11. Markussen JM (1984) Commercial exploitation of polymetallic nodules-when will it start? Fridtjof Nansen Foundation Newsletter No 1
12. NACOA (1983) Marine minerals: an alternative mineral supply. National Advisory Committee on Oceans and Atmospheres. Washington DC, p 53
13. Lipton I, Nimmo, M, Parianos J (2016) TOML Clarion-Clipperton zone project, Pacific Ocean. NI43–101 Report. Brisbane Australia, AMC Consultants Pty Ltd. Available at www.sedar.com
14. Padan JW (1990) Commercial recovery of deep seabed manganese nodules: twenty years of accomplishments. Mar Min 9:87–103
15. Siapno W (1985) Oral Presentation. Underwater Mining Institute
16. Siapno W (1990) Oral presentation. Underwater Mining Institute
17. UN (1986) Methodologies for Assessing the Impact of deep seabed minerals on the world economy. (E 86.11.A.13) Dept. of International, Economic and Social Affairs, New York, p 153
18. Earney FCF (1990) Marine mineral resources. Routledge, p 387
19. Herrouin G, Lenoble JP et al (1989) A manganese nodule industrial venture would be profitable: summary of a 4 years study in France. In: Offshore technology conference, p 321–332
20. Moore JR (1983) Marine hard minerals resources-progress and problems. Proc Oceans 83(111):1145–1149
21. Halbach P, Manheim FT (1984) Potential of cobalt and other metals in ferromanganese crusts on seamounts of the Central Pacific Basin. Mar Min 4:319–336

22. Clark AL, Humphrey P et al (1985) Cobalt-rich Crust Potential. OCS Study, MMS 85–006. US Dept. of the Interior, Minerals Management Service, p 35
23. Manheim FT (1986) Marine cobalt resources. Science 232:600–608
24. Wiltshire J (1989) Precious metal accumulation in manganese crusts from the Hawaiian and Johnson Island exclusive economic zones. In: Proc. 2nd Symp. Cont. Margins. University of Texas, p 15
25. Wiltshire J (1997) The world cobalt market and its ability to support manganese crust mining. In: Saxena N (ed) Recent advances in marine science and technology, vol 1997. Pacon International, pp 337–347
26. Odunton NA (2011) The international seabed authority: its roles organs and functions. Presentation Seminar on the work of the authority 28–30 March 2011, Kingston Jamaica. https://www.isa.org.jm/files/documents/EN/Seminars/2011/ISARoleFunctions-NAOdunton
27. Crockett RN, Chapman GR, Jones ALP (1984) British Geological Survey: mineral intelligence, statistics and economics, Rept. No 2/52

Epilogue

Keywords Nadir, Critical metals, International Seabed Authority, SOLWARA, Uncertainty

Deep-sea minerals were at their nadir at the start of the twenty-first century. They almost fell below the horizon of public consciousness. However, interest in them revived as the first decade progressed. There were several reasons for this:

(i) First, and most importantly, the prices of the commercially valuable metals in the nodules started to increase (Figure 1 of "Economic and Legal Developments, 1983–2000") after a long period of quiescence, during which terrestrial mines for the elements concerned had not been working at full capacity.

(ii) Metals requirements expanded from the Ni, Co, Cu, Zn and Mn which had been of principal interest in deep-sea minerals in the twentieth century to a host of rare and critical metals needed to support the green and technology industries that were newly emerging in the early years of the twenty-first century [1]. By the second decade of the century, the EU had classified Co, Ga, REE, In, Mg, PGE, Sb, and W as critical metals, of which only Co and REE (Fig. 1) have significant deep-sea sources. These metals were required in the manufacturing of new products such as mobile phones, special batteries, electric motors, high-efficiency bulbs, wind turbines, solar panels, and other things.

(iii) The International Seabed Authority started to open up for exploration the reserved areas granted to it under UNCLOS III and attempted to stimulate interest in deep-sea minerals in many other ways [2].

As a result of these factors, first, manganese nodules became of interest again, next polymetallic sulfides, and phosphorite third. Only Co-rich crusts seemed not to benefit from the twenty-first-century deep-sea minerals upturn.

From the start of the twenty-first century, the ISA became the dominant force in deep-sea minerals developments. According to Lipton et al [3], by 2011 interest in them had led the ISA to agree that preparations should start on the formulation of a

D. S. Cronan, *Deep-Sea Minerals Developments in the 20th Century*, https://doi.org/10.1007/978-3-031-52342-7

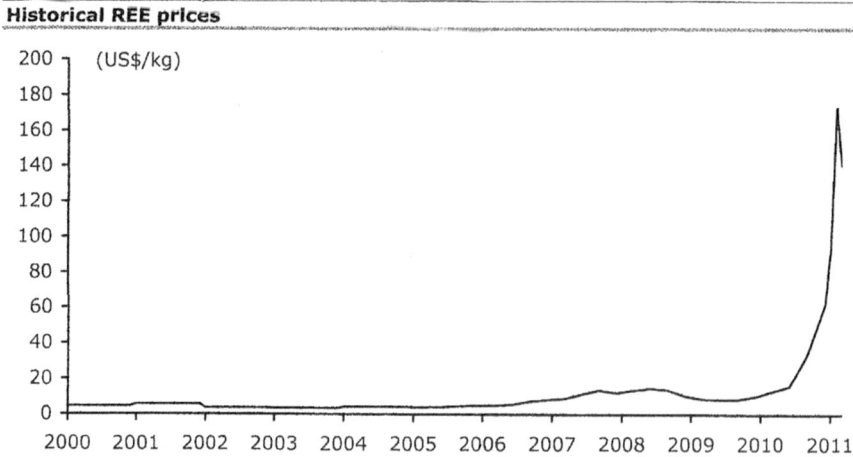

Fig. 1 REE prices in the first decade of the twenty-first century

mining code for the exploitation of deep-sea minerals in the International Seabed Area. This was in response to a significant increase in applications for exploration for both nodules and PMS in the Area, pursuant to exploration regulations already adopted by the ISA. Since its inception, the ISA had promoted the exchange of information, environmental science best practices, and technology regarding manganese nodules (and later PMS and Co-rich crusts), largely by means of workshops. Agendas, abstracts, presentations, and in some cases, videos of the presentations can be downloaded for many of these workshops (https://www.isa.org.jm/workshops). The ISA has also produced a series of over two dozen Technical Studies, Brochures, and Briefing Papers (https://www.isa.org.jm/documents-resources/publications). Over the years, the ISA has granted many licenses to explore for deep-sea minerals, the changing distributions of which can be viewed on the ISA website (www.isa.org.jm). At the time of writing it is in the process of drafting regulations under which any future deep-sea mining will take place [2].

The first revival of interest in manganese nodules was announced at the 2006 UMI by Germany. According to Kudrass et al [4], due to the strong increase in metal prices after 2002, the German Geological Survey (BGR) was charged by the German Federal Government with applying for a license area for exploration for nodules in the CCZ. This was granted in 2006. The new contract was in large part based on work done by Germany in the CCZ in the 1970s. In July 2011, Nauru Ocean Resources (NORI) and its contractor Deep Green Resources were granted exploration licences in four areas relinquished to the ISA under the UNCLOS III provisions. This was an important move as it paved the way for other developing states to apply for exploration licenses in the ISA-reserved areas, including Tonga, Kiribati, and the Cook Islands. The Cook Islands also revived interest in the nodules in their own EEZ. This had been quiescent since the 1990s, but the rising metals prices prompted them to look again at the issue [5]. In 2009, the Cook Islands passed the

Seabed Minerals Act, dedicated to the sensible, environmentally sound, development of its seabed mineral resources. This process was administered by the Cook Islands Seabed Minerals Authority which was established in June 2012 [6]. Consequent on this, mining companies were invited to tender for exploration licenses, and active exploration and evaluation of Cook Islands EEZ nodules is ongoing at the time of writing (Parianos Pers. Com. 2023). Other Pacific island states subsequently sought to re-evaluate their EEZ nodules [7]. A spin-off from the Cook Islands EEZ nodule exploration was the discovery of REE-rich sediments in the Cook Islands EEZ. Such sediments were first described Pacific-wide by Kato et al [8] and those in the Cook Islands EEZ have been described by Felix [9]. REE concentrations of up to almost 3000 ppm were noted, similar to or greater than those in manganese nodules in the same area. Finally, in a departure from twentieth-century thinking on hydrothermal manganese-rich crusts, Hein and Whitman [10] suggested that these deposits might be a source of battery-grade manganese and other critical metals.

Apart from SOLWARA 1, there was not much mining company interest in PMS deposits in the early years of the twenty-first century. SOLWARA 1 is a base metal deposit containing Cu, Zn, Au, and Ag, situated in about 1600 m water depth about 50 km from the port of Rabaul in PNG. In December 2009, Nautilus Minerals received a permit from the PNG Government to develop the prospect, with mining scheduled to start in around 2013. In the event, this did not happen. The project received much criticism on environmental grounds and was terminated near the end of the second decade of the twenty-first century for a variety of reasons. This operation was the furthest that any company had gone to develop a hydrothermal PMS deposit outside of the Red Sea. In the Red Sea itself, the development of the Atlantis II Deep had been on hold since the 1980s but towards the end of the first decade of the twenty-first century, Diamond Fields International developed an interest in reactivating the project. Subsequently, Manafai International of Saudi Arabia took up its development [11]. New work to help bring the deposit into production has also been carried out at the King Abdulla University of Science and Technology (KAUST) [12]. Elsewhere, there was interest in exploration for PMS deposits on the Central Indian Ocean Ridge [2], and on the mid-Atlantic Ridge [13] although it was felt that the latter would have a limited impact on the global metals market.

Marine phosphorite has not received much attention in the twenty-first century. Marine phosphorite mining may start if some phosphate-poor agricultural countries consider mining local offshore phosphorite in their EEZ rather than importing it eg. New Zealand. As described in Chapter "Phosphorites", there were attempts to develop the Chatham Rise phosphorites off NZ in the 1980s. There was another move to develop the Chatham Rise phosphorite deposits in the first and second decades of the twenty-first century, (https://www.rockphosphate.co.nz/the-project/) but this failed in 2015 and none has been mined to date. There has been a suggestion that the phosphorite off SW Africa might sometime in the future be of economic importance [14].

Although there has been little or no mining company interest in Co-rich crusts so far in the twenty-first century, surveys of them have continued by Universities and

Government Agencies. Hein et al. [15] suggested that any future mining of them would best take place around the summit regions of large guyots on flat or shallow inclined surfaces. Jahn and Halbach [16] also considered depth to be important in any future crust mining operations, with an optimum depth of between 800 and 2500 m.

In the third decade of the twenty-first century, the future for deep-sea minerals remains uncertain. Increasing concern about the effects of deep-sea mining on the marine environment and marine life [17, 18] has led some authorities to propose a moratorium on it. The earliest deep-sea mining environmental impact studies were reportedly conducted on the Blake Plateau in the early 1970s in conjunction with the Deep Sea Ventures operation there (Chapter "Activities on Manganese Nodules During the Post-war Boom"). Interestingly, towards the end of the second decade of the twenty-first century, the area was revisited by the US Bureau of Ocean Energy Management to assess its condition [19]. This work is ongoing at the time of writing and hopefully will help to provide a really long-term assessment of the likely environmental impacts of any future deep-sea mining. So far, none of the three marine minerals that have received mining company interest in the past as possible resources, nodules, PMS, or phosphorites, have been mined, and whether or not they might be mined in the future is unresolved.

References

1. Koschinsky A, Hein J et al (2010) Rare and valuable metals for high-tech applications found in marine ferromanganese nodules and crusts: relationships to genetic endmembers. Underwater Mining Institute
2. Jauhari P (2018) Deep seabed mining and the International Seabed Authority. In: Marine minerals: a new resource for the 21st century? Abs, Geol. Soc. Lond. Conf, p 63
3. Lipton I, Nimmo, M, Parianos J (2016) TOML Clarion-Clipperton zone project, Pacific Ocean. NI43-101 Report. Brisbane Australia, AMC Consultants Pty Ltd. Available at www.sedar.com.
4. Kudrass HR, Wiedicke M et al (2006) Polymetallic nodules as a future deep-sea mineral resource. Underwater Mining Institute
5. Cronan DS (2013) The distribution, abundance, composition, and resource potential of manganese nodules in the Cook Islands EEZ. Cook Islands Seabed Minerals Authority, Tech, Rept, No1, Rarotonga, Cook Islands
6. Anon (2015) Seabed mining and blue growth: exploring the potential of marine mineral resources. J Ocean Technol 10:61
7. Cronan DS (2018) Recent investigations on manganese nodules in the central Equatorial Pacific. In Marine minerals: a new resource for the 21st century? Abs, Geol. Soc. Lond. Conf, p 68
8. Kato Y et al (2011) Nat Geosci 4:535–539
9. Felix D (2018) Cook Islands deep-sea sediment-a possible rare earth resource. In Marine minerals: a new resource for the 21st century? Abs, Geol. Soc. Lond. Conf, p 26
10. Hein JR, Whitman S (2018) Hydrothermal manganese deposits from the global ocean as a resource for battery-grade manganese and critical metals. In Marine minerals: a new resource for the 21st century? Abs, Geol. Soc. Lond. Conf, p 61
11. Hamer D (2018) Atlantis II deposit, Red Sea-world's largest sea-floor massive sulfide mineral resource. In Mineral resources at the Frontier. Abs. William Smith Meeting 2018, Geol. Soc. Lond
12. Smith J, Modenesi MCI et al (2018) Formation and characterization of metalliferous sediments in the Red Sea. In, Marine minerals: a new resource for the 21st century? Abs, Geol. Soc. Lond. Conf, p 49
13. Cherkashov G (2018) Russian exploration of seafloor massive sulfides: results and prospects. In Marine minerals: a new resource for the 21st century? Abs, Geol. Soc. Lond. Conf, p 59

14. Coles SKP, Wright CI et al (2002) The potential for environmentally sound develop-
    ment of marine deposits of potassic and phosphate minerals offshore Southern Africa. Mar
    Georesources Geotechnol 20:87–110
15. Hein JR, Usui A, Dunham R (2007) Overview of cobalt-rich ferromanganese crusts, sea-
    mounts, and the outlook for mining. Underwater Mining Institute
16. Jahn A, Halbach P (2017)Resource estimation of Co-rich ferromanganese crust deposits,
    bathymetric and economic considerations. Underwater Mining Institute
17. Weaver P (2018) Potential environmental aspects of deep-sea mining. In Marine minerals: a
    new resource for the 21st century? Abs, Geol. Soc. Lond. Conf, p 65
18. McKie R (2023) Deep-sea mining will destroy the ecosystem, scientists warn. The
    Observer, p 6–7
19. www.boem.gov/marine

Index

A
ANZUS Tripartite Program, 79
Atlantis II Deep, 50, 52, 119, 120, 139

B
Benthic Impact Experiment (BIE), 116–118

C
Cameras and television, 63–65, 67, 90, 108
Central Indian Basin, 101
Chatham Rise, 60, 61, 109, 139
Clarion-Clipperton Zone (CCZ), 5, 6, 10, 11,
    14, 20–24, 27, 32, 47, 63, 69–71, 75,
    76, 96–99, 101, 102, 104, 108, 114,
    116, 125–127, 138
Committee for Co-ordination of Joint
    Prospecting for Mineral Resources in
    South Pacific Offshore Areas
    (CCOP/SOPAC), 24, 25, 54, 79, 80,
    82, 85, 88, 89, 102, 103
Cost models, 36–38
Critical metals, 137, 139

D
Dispersion haloes, 67
Disturbance, 11, 68, 114–119

E
East Pacific Rise, 23
East-West Centre, 104, 129

EEZs, 24, 56, 75, 76, 78, 79, 81, 85–92,
    99, 102–104, 108, 125, 131,
    138, 139
Environment, xx, 6, 11, 13, 23, 25, 27, 49, 68,
    96, 98, 113–120, 133, 140
Environmental Impact Statement (EIS),
    11, 91, 131
Epaulard, 22, 108

F
Fertilizers, 59, 61
Francolite, 59
Free-fall sampling, 1, 24

H
Hawaiian Archipelago, 43, 76, 131
Hawaii EEZ, 92
Hydraulic grab, 65, 109
Hydrothermal ferromanganese oxide
    crusts, 55–56
Hydrothermal plumes, 55, 67, 107, 110

I
Implementing Agreement (IA), 11, 14, 133
International Seabed Authority
    (ISA), 12–14, 98, 101, 127, 128,
    133, 137, 138

J
Jade Hydrothermal Field, 81

L

Law of the Sea, 7, 9, 11–14, 24, 35, 37, 69, 71,
    75, 76, 85, 95, 98, 99, 123,
    126–128, 132–134

M

Manganese nodules, 1, 3–12, 17–25, 27,
    30–39, 46, 63, 64, 66–72, 75, 76, 79,
    87, 89, 95–104, 107, 108, 114,
    116–119, 125–130, 137–139
Metalliferous sediments, 50, 52, 54–55,
    65, 78, 119
Metal Mining Agency of Japan (MMAJ), 22,
    79, 82, 87, 89, 102, 103, 108, 116
Mid-Pacific Mountains (MPMs), 27,
    44, 87, 88
Minerals Management Service
    (MMS), 91, 128

N

Nadir, 137
Nodule collectors, 13, 69, 114
Nodule mining Consortia, 96
Nodule mining tests, 69

P

Pacific Ocean, 5, 17–30, 44, 52, 75, 77,
    78, 85–92
Permissive areas, 46
Pioneer Investors (PIs), 14, 98, 127
Plumes, 10, 11, 55, 67, 107, 110,
    114–116, 118–120
Polymetallic sulfides (PMS), 52–56, 67,
    76–83, 95, 107–110, 119, 120, 124,
    125, 137–140

Post-war boom, 6, 17–39, 43, 71, 101, 102,
    115, 140
Preparatory Commission (Prep Com), 14,
    101, 127

R

Reciprocating States Regime
    (RSR), 127
Recolonization, 11, 114, 115, 117–118
Redeposition, 114, 116, 117
Red Sea brines, 50, 51, 65
Reserves and resources, 33, 34
Rock drill, 107, 109

S

Seamark, 108
Seamounts, 23, 44, 46, 47, 59, 60, 78, 87, 88,
    90, 91, 104, 130
SOLWARA, 82, 139
Sonne Hydrothermal Field, 79
Strategic metals, 38, 45, 75, 97

T

TAG Hydrothermal Field, 78, 82
Transects, 25–30, 100, 118

U

Uncertainty, 9–10, 35–38, 124, 126, 128
Upwelling, 60, 119
US Trust Territories, 76, 87

V

Voisey Bay, 130, 132, 133